Basic Infrared
Spectroscopy

Basic Infrared Spectroscopy

by J. H. van der Maas

Analytisch Chemisch Laboratorium der Rijkuniversiteit Utrecht

SECOND EDITION

HEYDEN & SON LTD

London · New York · Rheine

Heyden & Son Ltd., Spectrum House, Alderton Crescent, London NW 3XX
Heyden & Son Inc., 225 Park Avenue, New York, N.Y. 10017, U.S.A.
Heyden & Son GmbH, Steinfurter Strasse 45, 4440 Rheine/Westf., Germany

Library of Congress Catalog Card No. 70–101090
ISBN 0 85501 029 0   (paperback)
ISBN 0 85501 031 2   (cloth)

Made and printed in Great Britain at The Pitman Press, Bath

Contents

 # Foreword

Infrared spectroscopy is one of the branches of the physical sciences where practice ran far and wide, leaving theory struggling behind. Making sense of the spectra of large molecules involves theoretical methods often different from those which allow satisfactory treatment of the spectra of small molecules, and I believe that in spite of the venerable age of analytical infrared spectroscopy (a quarter of a century) new fruitful approaches will be found. This book, though not breaking new ground in infrared science, is an attempt to bring an understanding of the methods one uses in a reasoned approach to the interpretation of the spectra of large molecules to the novice in the field of analytical infrared spectroscopy; the author has attempted to restrict the subject matter to the necessary elements of this fascinating branch of science, but to give these a full treatment. May the quality of the work done by those who have used this book be his reward.

G. DIJKSTRA

Preface

The second edition of this book is only slightly different from the first one. Suggestions and critical remarks have been taken into account, while on request a general bibliography has been added containing several monographs devoted to different applications of infrared spectroscopy.

Appendix B has been enlarged with a few more typical band contours and explanatory text has been added.

With respect to the correlation tables a few remarks may be necessary. In general, present-day tables and charts fail for three reasons mainly:

(i) all published data are used without attention to the difference in accuracy. The regions therefore become broader and broader, while simultaneously the fine structure in a certain group gets lost.

(ii) in preparing the tables no allowance is made for the various sampling techniques, though it is well known that significant frequency shifts occur in different solvents.

(iii) sometimes data from the literature quoted are inaccurate or false, probably initially by accident but persisting thereafter due to carelessness.

The remedy seems to be simple but laborious: new charts and tables are to be prepared; e.g. tables for gases, for pure liquids, for solids in KBr, etc. The accuracy should be known. We believe this to be a necessary operation for there is a real danger that infrared spectroscopy will become ineffective as a tool for structure determination. Tables as such are in preparation but not yet ready. For the time being the less invaluable ones are to be used.

I would like to express my gratitude to Dr. M. A. Ford of Perkin-Elmer, Beaconsfield, for his interest and stimulating remarks and to Mr. E. T. G. Lutz for his help in preparing the additional spectra for this edition.

The co-operation of the publisher is gratefully acknowledged. As before, remarks and suggestions to improve the usefulness of this book will be very much appreciated.

THE AUTHOR

Units and abbreviations

Variable	Symbol	Unit	Unit Abbreviation
length	l	metre	m
time	t	second	s
mass	m	gram	g
temperature	T	degree	K or °C
energy	E	erg or joule	erg or J
intensity	I	joules per second	$J\ s^{-1}$
frequency	ν	hertz	Hz
wavelength	λ	micrometre	μm
wavenumber	σ	reciprocal centimetres	cm^{-1}
index of refraction	n	–	–
transmission	T	percent transmission	%T

Prefixes

Prefix	Abbreviation	Value
mega	M	10^6
kilo	k	10^3
milli	m	10^{-3}
micro	μ	10^{-6}
nano	n	10^{-9}
pico	p	10^{-12}

Conversion Factors and Constants

light velocity in vacuo $(c) = 3 \times 10^8$ m s^{-1}

E_{kinetic} at . . . K corresponds to	E_{photon} at . . . μm
100	144
300	48
1000	14·4

E_{photon} at 1 μm = 28·7 kcal/mole

E_{photon} at 1 nm $\equiv 2 \cdot 87 \times 10^7$ cal/mole

or $\equiv 1 \cdot 99 \times 10^{-16}$ J/mole

1 Å (ångström) = 100 pm $\equiv 0 \cdot 3 \times 10^{19}$ Hz

1
Introduction

ELECTROMAGNETIC RADIATION

Electromagnetic radiation extends from γ- and cosmic rays to radio frequencies and includes the ultraviolet, visible and infrared regions. It has been proved to be a periodically changing or oscillating electric field in a certain direction with a magnetic field oscillating at the same frequency but perpendicular to the electric field. One cannot produce one without the other (Fig. 1). The magnetic field will not concern us here.

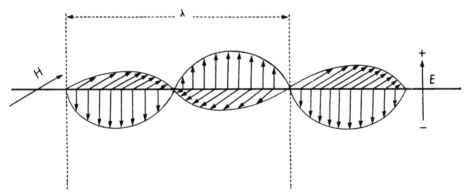

FIG. 1. Electromagnetic wave. H magnetic field, E electric field, λ wavelength

Electromagnetic radiation may be considered as a wave motion or as a stream of particles, often called quanta or photons. As a wave motion it can be characterised by a few parameters: the length of a wave, the wavelength (λ); the speed at which the wave moves, i.e. the velocity (c); and the frequency (v) being the number of waves or cycles per second. As for sound waves, the velocity for electromagnetic waves proves to be a constant for the medium in which the waves are propagating ($c = 3 \times 10^8$ m s^{-1} *in vacuo*).

It can be seen easily that

$$\lambda v = c$$

the wavelength being inversely proportional to the frequency. Turning back to the corpuscular or quantum character of electromagnetic radiation we have from quantum mechanics the relation

$$E = hv$$

(E = energy per quantum of radiation, h = Planck's constant). Thus ν is related linearly to the energy of the radiation.

The electromagnetic spectrum can be divided into several regions (Fig. 2) differing in frequency only, but each with its own special character. The static field corresponds

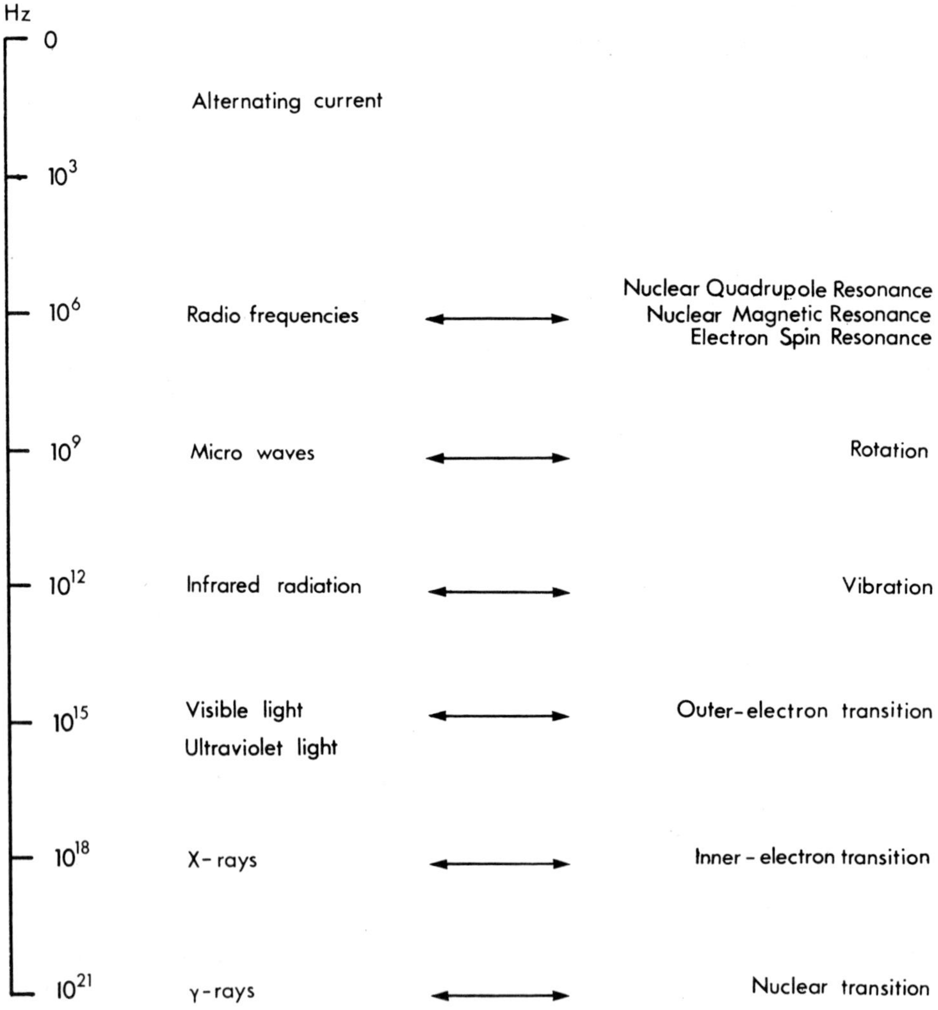

Hz

0			
	Alternating current		
10^3			
10^6	Radio frequencies	⟷	Nuclear Quadrupole Resonance Nuclear Magnetic Resonance Electron Spin Resonance
10^9	Micro waves	⟷	Rotation
10^{12}	Infrared radiation	⟷	Vibration
10^{15}	Visible light Ultraviolet light	⟷	Outer-electron transition
10^{18}	X-rays	⟷	Inner-electron transition
10^{21}	γ-rays	⟷	Nuclear transition

FIG. 2. Electromagnetic spectrum. The right hand side lists some possible spectroscopic sources and absorptions

to $\nu = 0$. The highest frequencies are found at about 10^{20} Hz. Infrared radiation is to be found at frequencies of 10^{14}–10^{12} Hz, corresponding to wavelengths of 1–100 μm or wavenumbers of 10 000–100 cm^{-1}.

At room temperature (300 K) the average velocity of a gas molecule is about 500 m/sec. So in 10^{-7} s a molecule moves 50 μm. In the same time a radio wave with a frequency of 10^7 Hz completes one cycle. This means that the alterations in the electric field are followed easily by the relatively fast-moving molecules; seen from the centre of gravity of the molecule the radio frequencies seem to be more or less 'static'. At frequencies of about 10^{12} Hz, however, it is just the other way round: the molecule seems to be static with respect to the fast-oscillating electric field. Interaction of electromagnetic radiation and molecules will be possible in two ways: with or without energy exchange.

REFRACTION AND DISPERSION

Apart from the rotation of polarised light, refraction (or dispersion) is one of the few mechanisms of interaction between molecules and electromagnetic radiation without exchange of energy. It is the result of different velocities in different media. The index of refraction is given by

$$n = \frac{\text{velocity of light } in \ vacuo}{\text{velocity of light in a medium}}$$

This index n is not a constant but depends to some extent on the wavelength of the radiation. In general n and $dn/d\lambda$ decrease in magnitude as λ increases (Fig. 3). It

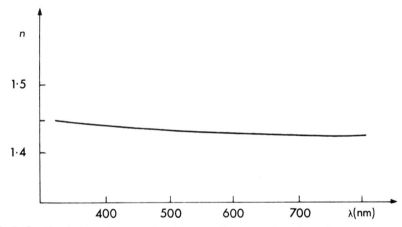

FIG. 3. Refractive index of a non-absorbing medium as a function of the wavelength. The line can be represented by the Cauchy (1836) expression: $n = A + B/\lambda^2 + C/\lambda^4$, where A, B and C are constants characteristic for the medium concerned

is found, however, that as one approaches and passes wavelengths at which the medium absorbs, the index of diffraction as well as $dn/d\lambda$ are subject to great alterations (Fig. 4). This is often referred to as anomalous dispersion. Unfortunately, one cannot make use of this high dispersive power because absorption increases too, and thus too much radiation will be absorbed if prisms of normal dimensions are used. It is worth noting, however, that the highest dispersion is found near to an absorption band at its 'long-wavelength side'.

ABSORPTION

Absorption is a form of interaction between matter and light or radiation in which energy exchange is involved. We will restrict ourself to the absorption of infrared radiation by molecules. Under what condition(s) can absorption take place?

As we know already, infrared radiation is a moving oscillating electric field, oscillating at frequencies of 10^{12}–10^{14} Hz. The internal movements of the atoms in a molecule with respect to its centre of mass occur at the same frequency. Let us consider a mechanical model first. Suppose one has a ball connected to the ceiling by an

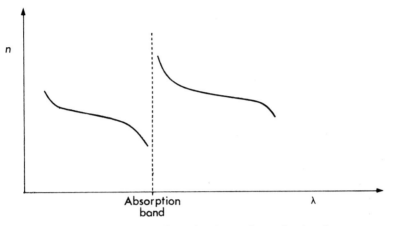

FIG. 4. Anomalous dispersion in an absorption band

elastic rubber string and suppose the ball is moving up and down. The ball will naturally come to rest sooner or later due to the gravitation force and the friction with the air. Now to keep it on moving one has to deliver energy to the system. This can be done by flipping the ball *gently* at the moment it starts moving upwards. In other words, one has to supply energy at *the same frequency* at which the system is moving. If

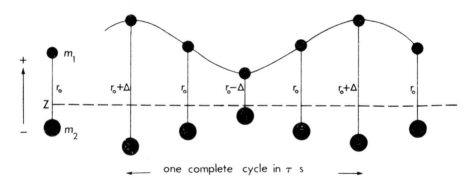

FIG. 5. Vibration of a diatomic molecule. $m_2 = 2m_1$; z, centre of mass; r_0, distance between m_1 and m_2 in the equilibrium position

this is done, the maximum distance from ball to ceiling will steadily increase. If one just touches the ball or hits it at another frequency, the system will soon come to rest.

Now consider a diatomic molecule with masses m_1 and m_2 at an equilibrium distance r_0 (Fig. 5). If such a molecule is vibrating, the masses m_1 and m_2 are moving towards and away from each other, i.e. the distance r is subject to periodic changes. The centre of mass of the molecule is considered to be fixed, otherwise translation would be involved. To keep this centre at one place, atom 1 must move twice as much as atom 2, due to the difference in mass. Assuming a difference in electric charge between m_1 and m_2, a dipole exists for this molecule, directed for instance as shown in the diagram. During the vibration of the molecule the dipole moment will change simultaneously with the distance r. As this change is periodical, such a vibrating molecule produces a stationary alternating electric field, the frequency being $1/\tau$ Hz. As in the mechanical model this system can absorb energy provided it is supplied in a similar way, i.e. by a moving electromagnetic field oscillating at the same frequency $1/\tau$ as that of the molecule itself.

If, however, this molecule has no alternating dipole, no oscillating electric field would arise and therefore no energy could be absorbed from any external source, in spite of the fact that this electromagnetic radiation might be of the correct frequency.

SELECTION RULES

The vibration of the diatomic molecule is a very simple one. For polyatomic molecules many vibrations exist, as we will see later. Nevertheless, what holds for the vibration of the diatomic molecule holds for any vibration in any molecule. This enables us to state some rules:

1. Absorption of infrared radiation by a vibrating molecule will only take place if the vibration produces an alternating electric field; i.e. if the vibration is coupled with a changing dipole moment.

2. In order for the radiation to be absorbed, the vibration frequency of the molecule must be identical to the frequency of the radiation, and furthermore since $E = h\nu$, the absorbed quantum of energy will have a distinct value. In other words the absorption of energy is quantised.

DISSIPATION OF ABSORBED ENERGY

The energy absorbed by a molecule is rapidly dissipated. The excited molecule loses its energy (vibrational and rotational) in less than 10^{-6} s. The energy is either transformed into kinetic energy as result of collisions or released again as a photon. As the direction of the liberated photon is random in space, and because the absorption process can be repeated for such a photon on its way through the medium, it can be seen that for a once-absorbed photon the probability of re-emerging from the medium in the direction of the transmitted beam is negligibly small.

BIBLIOGRAPHY
W. Brügel, *An Introduction to Infrared Spectroscopy*, Methuen, London, 1962.
R. P. Bauman, *Absorption Spectroscopy*, John Wiley and Sons, New York, 1962.

2
Theory

Any movement of an atom in space can be represented by 3 independent mutually perpendicular movements parallel to the x, y and z axes in a cartesian system. The atom is said to have 3 degrees of freedom. A system of N free-moving atoms thus will have $3N$ degrees of freedom.

Like an atom, the movement of any object in space, or rather the movement of its centre of mass, also called translation, can be described by 3 parameters. In addition, the object may rotate about the centre of mass. This rotation can also be represented by 3 parameters, i.e. three rotations on three cartesian axes originating in the centre of mass.

It follows that to describe the movements of an object in space 6 degrees of freedom are required: three for translation and three for rotation. The same holds for a molecule of N atoms. For rotational and translational movements of the molecule 6 degrees of freedom are required, thus leaving $3N-6$ degrees for the movement of the atoms in the molecule, i.e. for the vibrations of the molecule.

VIBRATION

The movements of the atoms in a molecule may be very complex but as was shown above these complex vibrations can be composed from $3N-6$ basic vibrations, the so-called 'normal' or 'fundamental' vibrations. Provided these $3N-6$ vibrations satisfy the earlier stated selection rules, the molecule will give rise to at least $3N-6$ absorption bands somewhere in the infrared region. For strictly linear molecules, such as carbon dioxide, the number of normal vibrations is $3N-5$. Consider the three rotational axes of such a molecule. The position of one or more atoms will be changed if rotation about two of these axes is carried out. Rotation about the third axis, the one which coincides with the molecular axis, does not change the position of the atoms. Hence only 2 parameters – two degrees of freedom – are required to describe any rotation of such molecules, so leaving $3N-5$ degrees of freedom for vibrational analysis.

Diatomic molecules

The number of normal or fundamental vibrations for this type of molecule is just one. According to the selection rules this vibration will only give rise to absorption of radiation in cases where the atoms are different, e.g. HCl, CO, ICl, NO etc. For only then can a dipole moment exist. The symmetric molecules such as H_2, N_2, O_2, Cl_2 etc. will not absorb, as there will be no changing dipole moment.

Harmonic oscillator. At what frequency v will a diatomic molecule have its normal vibration? Consider the earlier-mentioned mechanical model: two masses m_1 and m_2 connected by a spring at a distance r_0. If this distance has to be increased or decreased by Δr a force F has to be applied which is proportional to Δr (Hooke's law for a harmonic oscillator):

$$F = -f \Delta r$$

where f is the proportionality factor or force constant. This holds as long as Δr is reasonably small. So if this system is oscillating, Δr being small, a simple harmonic motion will be the result. The frequency of such an oscillation is known to be given by:

$$v = \frac{1}{2\pi} \cdot \sqrt{\frac{f}{\mu}}$$

where μ is the reduced mass, to be found from

$$\mu = \frac{m_1 m_2}{m_1 + m_2} \quad \text{or} \quad \frac{1}{\mu} = \frac{1}{m_1} + \frac{1}{m_2}$$

As can be seen, the frequency of a harmonic oscillator depends only on the force constant f and the reduced mass μ (see Table 1).

TABLE 1[a]

Molecule	μ (1·66 × 10^{-24} g)	f (10^5 dyne/cm) (N cm^{-1})	$\sigma(v)$ (cm^{-1})	$2B_0$ (cm^{-1})
H$_2$	0·50	5·07	4160[b]	61
HD	0·67	5·15	3631[b]	46
D$_2$	1·00	5·24	2993[b]	30
^{35}Cl$_2$	17·50	3·21	556[b]	0·3
N$_2$	7·00	22·2	2331[b]	2·0
O$_2$	8·00	11·3	1555[b]	1·4
HF	0·95	8·62	3935[c]	21
H^{35}Cl	0·97	4·74	2886	10·6
HBr	0·98	3·78	2558	8·5
HI	0·99	2·89	2233	6·6
NO	7·46	15·4	1877	1·7
CO	6·85	18·6	2143	2

[a] The data are taken from K. W. F. Kohlrausch, *Der Smekal-Raman-Effect*, and from G. Herzberg, *Molecular Spectra and Molecular Structure*. (See bibliography for this chapter.)
[b] Taken from Raman spectroscopy as these vibrations are infrared inactive.
[c] Extrapolated value.

Energy. What is the total energy of such a classical harmonic oscillator? Suppose there will not be any friction losses due to any kind of interaction. Once moving, therefore, the system will maintain a constant total energy E, being the sum of the varying kinetic and potential energies, T and V, i.e. $E = T + V$. According to Hooke's law the potential energy V is at any time:

$$V = \tfrac{1}{2}f(\Delta r)^2$$

A curve representing V as a function of Δr will be a parabola (Fig. 6). Suppose the distance r between m_1 and m_2 during the oscillation is varying from $r_0 + \Delta r_{max}$ to $r_0 - \Delta r_{max}$, Δr_{max} being the maximum displacement. When the distance is $r_0 + \Delta r_{max}$ or $r_0 - \Delta r_{max}$ the kinetic energy of the system is zero, there being no movement for an infinitely small moment; hence $E = \frac{1}{2}f(\Delta r_{max})^2$.

When the distance $m_1 - m_2$ is r_0 the potential energy is zero and the kinetic energy $T = \frac{1}{2}f(\Delta r_{max})^2$.

From the foregoing we may conclude that the total energy of a classical harmonic oscillator depends only on the force constant f and the maximum displacement Δr_{max}. Hence any E value is allowed, for Δr_{max} can take any value; E might therefore be thought to be continuously variable.

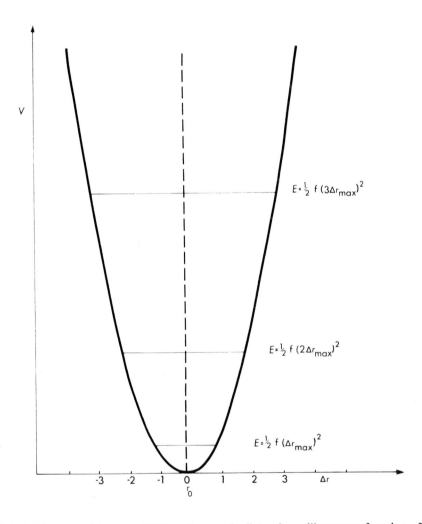

FIG. 6. The potential energy (V) for a harmonic diatomic oscillator as a function of Δr

From quantum-mechanical considerations however, one knows the energy values for a diatomic molecule are certainly not continuous. Equally spaced energy levels exist, as given by:

$$E_n = (n + \tfrac{1}{2})h\nu$$

where $n = 0, 1, 2, 3 \ldots$ (any positive integer). Two adjacent levels differ by an energy $h\nu$, while the lowest level E_0 is called the 'zero-point' energy. The quantum $h\nu$ is in agreement with what is said in the selection rule, although it seems to be in contradiction with what has been found for the mechanical model. ν for the mechanical

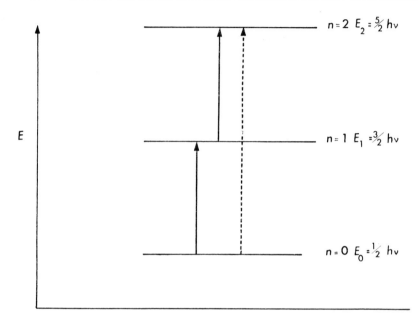

FIG. 7. Energy levels for a diatomic harmonic oscillator

model (masses in g) will be extremely small compared with ν of the molecule (masses in 10^{-23} g); in fact about 10^{11} times smaller if only the masses are considered. Since the energy is proportional to ν, the spacing between the energy levels in the case of the mechanical system will be so small that a continuum seems to be produced.

A molecule in the ground state E_0, absorbing an energy quantum $h\nu$, will reach the first excited state E_1. The same molecule may absorb again a quantum $h\nu$ and so reach the second excited state E_2. If, however, the molecule in the ground state absorbed a quantum $h\nu'$, where $\nu' = 2\nu$ it would have reached level E_2 at once (Fig. 7, dotted line). A condition from quantum mechanics states, however, that the only allowed transitions for a *harmonic* oscillator are those between adjacent levels or

$$\Delta E = E_{n+1} - E_n = h\nu$$

For a harmonic oscillator the transition

$$E_0 \xrightarrow{+h\nu'} E_2$$

is forbidden, while

$$E_0 \xrightarrow{+h\nu} E_1 \xrightarrow{+h\nu} E_2$$

is allowed. In the latter case both quanta have the same frequency and therefore the absorptions will occur at the same place in the infrared. Both absorptions will be indistinguishable. Furthermore, in order for a transition to occur between E_1 and E_2, a molecule which has just absorbed a quantum $h\nu$ must meet a second quantum before the gained energy has been dissipated. Since the number of excited (E_1) molecules is low according to Boltzmann's distribution law

$$\frac{N_1}{N_0} = \exp\left(\frac{-(E_1 - E_0)}{kT}\right)$$

the probability for a transition $E_1 \rightarrow E_2$ will be low. N_1/N_0 is the ratio of molecules in level E_1 to those in level E_0, while $\Delta E = h\nu = hc\sigma$. At room temperature ($\sim$300 K) and for $\sigma = 900$ cm^{-1} this ratio will be about 1:77. For 600 and 1200 cm^{-1} this ratio

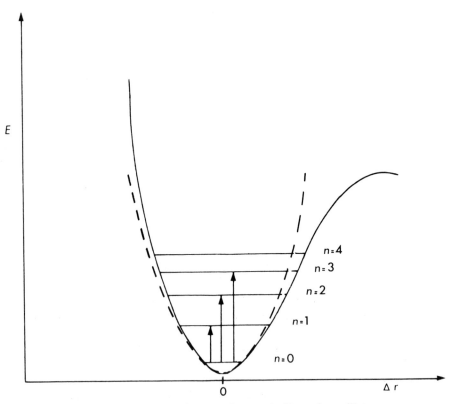

FIG. 8. Energy levels for an anharmonic diatomic oscillator.
- - - - - - parabola, ——————— potential energy curve

is 1:18 and 1:300 respectively. Thus spectra of doubly excited molecules may be observed at low frequencies.

Summarising the foregoing sections, we may conclude that a diatomic molecule acting as a harmonic oscillator appears to have but one transition. This is the basic or fundamental absorption, i.e. the one from the ground level E_0 into the first excited state E_1. In general these fundamentals fall between 4000 cm^{-1} and 400 cm^{-1}, the principal infrared region.

Anharmonic oscillator. In actual practice the vibration of the diatomic molecule is not strictly harmonic. The potential energy curve as a function of the displacement from the equilibrium distance r_0 is in fact not a parabola though it shows a close resemblance (Fig. 8). In the low-energy region the parabola is a good approximation to the true energy curve. The higher energy levels including the second are more closely approximated by

$$E_n = h\nu[(n + \tfrac{1}{2}) - x(n + \tfrac{1}{2})^2]$$

x being the anharmonicity constant, usually a small fraction of unity. The spacing between two levels is given by

$$E_n - E_{n-1} = \Delta E = h\nu(1 - 2nx) \quad n = 1, 2, 3 \ldots$$

and hence will differ from level to level. For the transition $E_0 \rightarrow E_1$ the energy difference will still be $h\nu$ provided x is much smaller than unity.

Overtones. As we have a harmonic oscillator no longer, transitions from $E_0 \rightarrow E_2$ or $E_0 \rightarrow E_3$ are allowed, but with much lower probability now than the fundamental. They do in fact occur. The vibrational frequency of the molecule is still ν, the ν that can be calculated from the equation on p. 7. The only difference between molecules in any of the vibrational levels is the amplitude of the mode, Δr. The transition $E_0 \rightarrow E_2$ is called the first overtone, $E_0 \rightarrow E_3$ the second overtone. Both are indicated in Fig. 8. Furthermore, the smaller the x-value, i.e. the closer the approximation to the pure harmonic oscillator, the smaller the probability. Since $\sqrt{I}$ is proportional to the probability of a transition (I being the intensity of an absorption band), the smaller the anharmonicity constant x, the smaller will be the absorption at the overtone frequencies compared with the fundamental. Conversely, intense overtones point to a large degree of anharmonicity of a vibration.

For positive values of x, $\Delta E_{01} > \Delta E_{12}$ and therefore $\sigma_{02} < 2\sigma_{01}$, i.e. the wavenumber of the first overtone is less than twice that of its fundamental, whereas for negative x values the opposite is true.

ROTATION

Rigid diatomic molecule

We start again with the diatomic molecule, masses m_1 and m_2 joined by a rigid bar this time, length r_0. Such a molecule may rotate about the 3 axes through the centre of mass Z. Z is defined by $m_1r_1 = m_2r_2$; see Fig. 9. Rotation about these axes produces

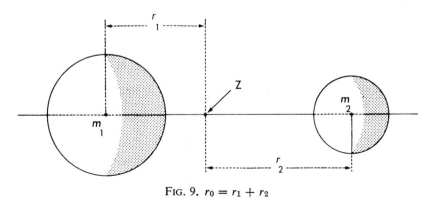

FIG. 9. $r_0 = r_1 + r_2$

three moments of inertia, I_a, I_b and I_c. The a axis coincides with the rigid bar (bond axis) and so the moment of inertia $I_a = 0$. Furthermore $I_b = I_c$ as the b and c axes are equal. I_b is defined as

$$I_b = m_1r_1{}^2 + m_2r_2{}^2$$

or, with $m_1r_1 = m_2r_2$ and μ, the previously mentioned reduced mass

$$I_b = \mu r_0{}^2$$

Now suppose the molecule has a permanent dipole moment directed to m_1, as indicated in Fig. 10. Viewed from the left to the right, the molecule rotating in the

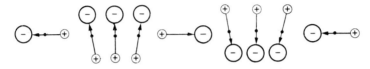

FIG. 10. Movement of a dipole during rotation

plane of the drawing produces an alternating dipole moment. Therefore such a molecule can absorb electromagnetic radiation provided its rotational frequency and that of the incident radiation are identical. Molecules without a permanent dipole moment cannot absorb radiation as they do not produce an alternating field during rotation.

Just as vibrational energy is quantised, so also is rotational energy. From quantum mechanics it can be shown that the rotational energy levels for the diatomic rigid molecule are given by:

$$E_J{}^{\text{rot}} = \frac{h^2}{8\pi^2 I_b} J(J + 1), \text{ where } J = 0, 1, 2 \ldots$$

J is called the rotational quantum number, while the selection rule is

$$\Delta J = \pm 1$$

The energy between two adjacent levels is thus given by

$$\Delta E_{J \to J+1}{}^{\text{rot}} = \frac{2h^2}{8\pi^2 I_b} \cdot (J + 1)$$

or in wavenumbers

$$\sigma_{J \to J+1}{}^{\text{rot}} = \frac{2h}{8\pi^2 I_b c} (J + 1)$$

and since for a molecule $h/8\pi^2 I_b c$ is constant the relation can be written as

$$\sigma_{J \to J+1}{}^{\text{rot}} = 2B(J + 1)$$

where B is called the rotational constant. The levels are shown in Fig. 11 in cm^{-1} (E/hc), a common and useful method as any transition can be read off and searched for in a spectrum directly this way. A presentation as such is called a 'term scheme'. As $\Delta J = \pm 1$ the absorption of radiation will take place at $2B$, $4B$, $6B$ cm^{-1} etc., and hence the distance between two adjacent absorption peaks will always be $2B$ cm^{-1}, i.e. the peaks are equally spaced (see Fig. 14) whereas the energy levels are not. For some molecules the value $\Delta\sigma_{0\to1}$ is given in Table 1. As one can see these absorptions occur in the far infrared region.

Absorption process. The absorption of electromagnetic radiation by a molecule and the conversion in rotational energy may be visualised as follows.

Consider a non-rotating molecule, the centre of mass being fixed anywhere in space (no kinetic energy). Let the molecule be irradiated by monochromatic radiation, or in other words by a stream of photons. Suppose a photon reaches the molecule. Since the time of interaction between a photon and the molecule is very short – about

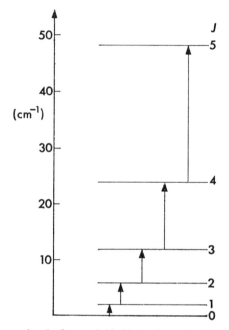

FIG. 11. Rotational energy levels for a rigid diatomic molecule. The arrows indicate the allowed absorption transitions

10^{-18} s – we will have to trace the photon on its way through the molecule at extremely short time intervals.

The photon is (creates) an alternating electric field and a molecule with a permanent dipole will start rotating in order to follow the alterations of the field (*cf.* electric motor: a coil in an alternating field). As long as the molecule rotates in phase with the alternating electric field it will withdraw energy from the photon and return it when moving out of phase.

The rotational energy of a molecule, given by $E = \tfrac{1}{2}I\omega^2$, is as we know quantised, and thus also the angular velocity of the rotation, ω, which is related directly to the frequency of the alternating electric field of the molecule itself.

Two possibilities arise now. Either the frequency of the photon is identical to the first of the quantised angular frequencies of the molecule or it is not. The former will give rise to a complete absorption of the photon by the molecule because it moves in phase all the time the photon is 'present'. The latter may look complicated but is rather simple. Two waves, slightly different in frequency, will be more or less in phase

during a certain period and out of phase for another (see Fig. 12). For the molecule (and the photon) this will lead to an absorption of energy in the first half-period, followed by an equal return in the second. The final result will be an undisturbed passing of the photon. In short, the molecule 'absorbs' the photon, tries to fit it into the first excited energy level and if it cannot, lets it pass again.

For a molecule that is rotating already, a similar model can be set up. When a photon reaches such a molecule it will try to absorb the photon by increasing its

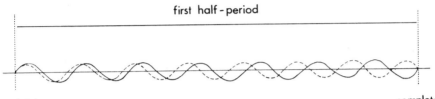

completely
in phase

completely
out of phase

FIG. 12. Phase relationships of two waves

angular frequency to the nearest quantised situation. If this frequency and that of the photon are the same, the photon will be absorbed. If not it will reject it for the same reason as mentioned before.

Non-rigid rotator

For a non-rigid bonding between m_1 and m_2, as is usually the case, the distance m_1-m_2 is no longer constant and hence the moment of inertia I_b is no longer constant either. The distance will be increased due to the centrifugal force, especially at high rotational velocities. The molecule may vibrate as well, but as the corresponding frequency is about a hundred times greater than its rotational frequency, one can make use of an average value for the varying distance r.

Nevertheless the spacing between the energy levels is no longer given by the earlier-mentioned relation and has to be changed to

$$\sigma_{J\to J+1}{}^{\text{rot}} = 2B(J+1) - 4D(J+1)^3*$$

where D is the centrifugal distortion constant, which is a positive quantity, roughly $B \times 10^{-3}$. The outcome can be seen in the term scheme (Fig. 13) and the corresponding absorption pattern (Fig. 14). As shown, the correction term proves to be without much effect for low J values.

Intensity of rotational bands

The intensity of a rotational band, belonging to the transition from level J to level $J+1$, is proportional to the number of molecules going over from $J \to J+1$. The probability is the same for each transition $J \to J+1$; therefore the intensity is proportional to the number of molecules in a certain level, in our case in level J. Here one can make use again of Boltzmann's distribution law:

$$\frac{N_J}{N_0} = \exp\left(\frac{-\Delta E_{0\to J}}{kT}\right)$$

$$\Delta E_{0\to J} = BJ(J+1)hc$$

* Calculated from $F_{\text{rot}} = E_{\text{rot}}/hc = BJ(J+1) - DJ^2(J+1)^2$.

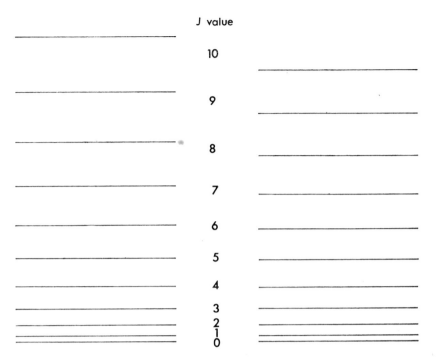

FIG. 13. Term scheme for the rotational energy levels for a rigid (left) and a non-rigid (right) rotator

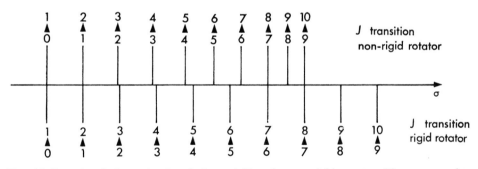

FIG. 14. Rotational absorption bands for a rigid and a non-rigid rotator. The wavenumber scale is arbitrary

where N_J is the number of molecules in level J and N_0 the number of molecules in the ground state, $J = 0$. Obviously N_J decreases rapidly with increasing J, as shown in Fig. 15.

Although we have always spoken about one level for each J it appears that each level has to be split into several completely identical levels; the number of these so-called degenerate levels is given by $2J + 1$. Hence the number of molecules in a certain level is proportional to the product of the ratio N_J/N_0 and the number of levels $2J + 1$. Thus the intensity of an absorption band is proportional to

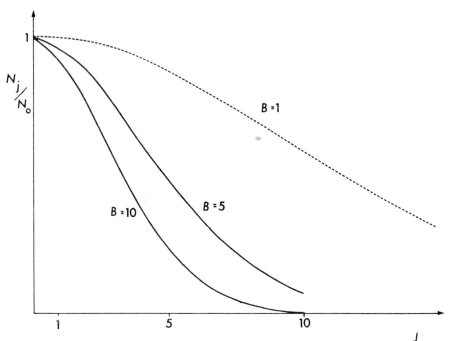

FIG. 15. The ratio N_J/N_0 for various values of B as a function of the rotational quantum number J, at room temperature

$N_J/N_0\,(2J+1)$ and, when this is plotted against J, curves are obtained as shown in Fig. 16 for certain B values. The peak with maximum intensity will be found at

$$J = \left(\frac{300}{2\cdot 88B}\right)^{\frac{1}{2}} - \tfrac{1}{2}$$

Intensity of vibrational bands

The intensity of a vibrational band belonging to the transition from $n = 0 \rightarrow n = 1$ proves to be proportional (by quantum mechanics) to the square of the change of the dipole moment for the corresponding normal vibration near the equilibrium position. In other words, for a linear diatomic molecule $I \approx \left(\dfrac{\partial M}{\partial x}\right)_0^2$. For more complex molecules I is proportional to

$$\left(\frac{\partial M_x}{\partial x}\right)_0^2 + \left(\frac{\partial M_y}{\partial y}\right)_0^2 + \left(\frac{\partial M_z}{\partial z}\right)_0^2$$

with M_x, M_y, M_z being the three components of the dipole moment M of the molecule in the x, y and z direction in the displaced position of the atoms. Unfortunately the derivative of M is not known beforehand and so one does not know anything about the intensity of a band. From practice, however, it appears that in many cases the partial dipole moment, M_{part}, can be used as a rough approximation to the derivative. For instance, a C=O group has a large M value and it absorbs strongly in the infrared, while the C≡C group in $CH_3CH_2C≡CCH_3$ shows hardly any absorption, in accordance with the expected small M. For more complex vibrations M is difficult to determine. Experience can then be of great help.

The intensity of overtones, $n = 0 \rightarrow n = 2$, which are only possible for anharmonic vibrations, depends on the anharmonicity term. The larger this factor the more intense the absorption band, though it is still much weaker than the fundamental

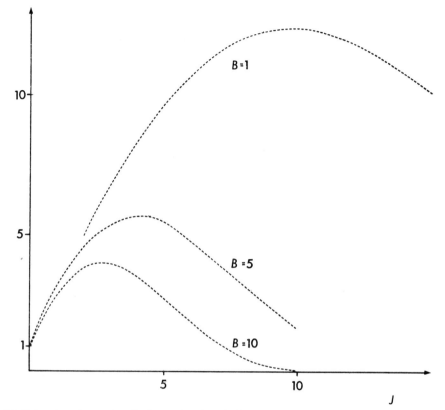

FIG. 16. Intensity of rotational absorption bands for various B values as a function of J. I in arbitrary units

absorption under identical sampling conditions. The CO group is a good example of a group producing an overtone (∼3400 cm⁻¹). Again one can make use of a rule of thumb: the stronger the fundamental the stronger the first overtone.

VIBRATING ROTATOR

From the preceding sections we know the energy transitions for a vibrating molecule to be about a hundred times greater than the rotational energies. Since these energies are so different we may, as a first approximation, consider that the molecule can vibrate and rotate independently. This assumption is a 'Born-Oppenheimer' like approximation, mathematically expressed as

$$E_{mol} = E_{vib} + E_{rot}$$

Substituting the expressions for E_{vib} and E_{rot} as found before one obtains

$$E_{mol} = h\nu_e[(n + \tfrac{1}{2}) - (n + \tfrac{1}{2})^2 x] + hc[BJ(J + 1) - DJ^2(J + 1)^2]$$

or in wavenumbers

$$F_{mol} = \sigma_e[(n + \tfrac{1}{2}) - (n + \tfrac{1}{2})^2 x] + BJ(J + 1) - DJ^2(J + 1)^2$$

where $F = E/hc$ and the subscript e denotes an anharmonicity-corrected frequency or wavenumber.

The term scheme for this expression is shown in Fig. 17 for $n = 0$ and for $n = 1$. It may be shown that the selection rules for the combined motions are the same as

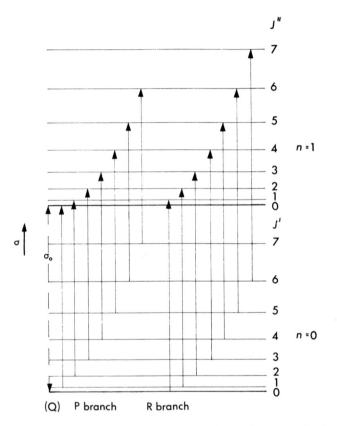

(Q)   P branch   R branch

FIG. 17. Term scheme for a diatomic anharmonic oscillator. Not to scale. In the P branch, $m = J + 1$, $m < 0$, $\Delta J = -1$; in the R branch, $m = J + 1$, $m > 0$, $\Delta J = 1$

those for each separately: $\Delta J = \pm 1$ and $\Delta n = \pm 1, 2 \ldots$ (for a diatomic molecule $\Delta J = 0$ is not allowed except under special circumstances).

The distance between two rotational levels, one belonging to $n = 0$ and one to $n = 1$ is given by

$$\sigma(J)_{0 \to 1} = \sigma_0 + 2Bm - 4Dm^3$$

$$\text{Where} \quad m = J + 1 \text{ and } m > 0 \text{ for } \Delta J = +1$$
$$m < 0 \text{ for } \Delta J = -1$$

$$m = \pm 1, \pm 2, \pm 3 \ldots$$

Thus absorption bands may be observed at the frequencies indicated in Fig. 18. Bands at the low frequency side ($\Delta J = -1$) are referred to as the P branch, while those to the high frequency side ($\Delta J = +1$) are called the R branch. The band for $\Delta J = 0$, uncommon for diatomic molecules, is called the Q branch.

According to the derived formula the spacing between two adjacent bands will be about $2B$ cm^{-1} for small m-values whereas it will decrease for increasing values of m (see Fig. 18a).

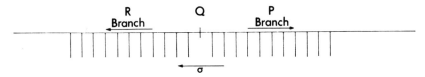

FIG. 18. Distribution of the rotational bands superimposed on a vibrational transition for a diatomic molecule

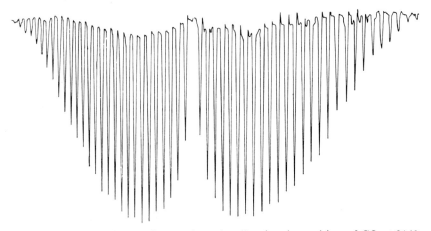

FIG. 18A. Rotational bands superimposed on the vibrational transition of CO at 2140 cm^{-1}

In fact the formula for the energy levels of a rotating vibrating diatomic molecule is somewhat more complex than the one which has been derived but we shall not go into more detail here.

Molecular interaction

Strictly speaking the formulae derived in the preceding sections are valid only for an isolated molecule. Actually the absorption of vibrational or rotational energy by a single particle cannot be observed. A much greater number of molecules is required, but some kind of interaction may then arise.

Gas phase. Interaction between molecules will be lowest in this phase especially at diminished pressure. At room temperature the molecules are moving in all directions and depending on the free path length they will collide regularly, e.g. for carbon monoxide about 5×10^9 times/second at 750 mm Hg and 300 K. At the moment of collision energy can be exchanged: one molecule may gain energy at the cost of another

one. The exchanged energy is either kinetic, or rotational, or vibrational or of combined character; see Fig. 19.

The conversion of vibrational and rotational energy into kinetic energy is one way
by which molecules dissipate their absorbed radiation. It may be observed as a rise in
temperature.

The conversion of kinetic energy (300 K) into vibrational energy is of no importance
at wavenumbers above 400 cm^{-1} as the energy is too small to reach the first vibrational
level. Since, however, the energy quanta required for rotational transitions are much

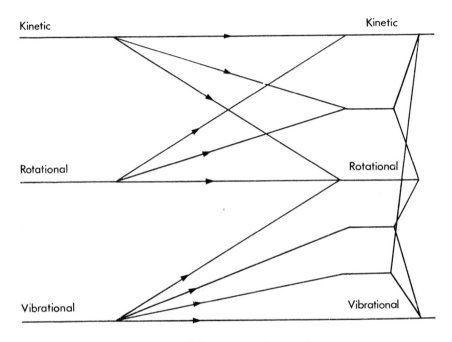

FIG. 19. Possible energy exchange scheme

smaller, these will be readily picked up from kinetic energy. At room temperature,
therefore, many molecules are rotating, i.e. they occupy rotational energy levels
($J \neq 0$). Consider a rotating CO molecule, rotational frequency v_J, at room temperature. As a result of kinetic energy, collisions with other molecules will occur every
2×10^{-10} s. Suppose a photon, frequency v_{J+1}, arrives just after a collision has
taken place. Let v_{J+1} be slightly smaller than $v_J + 2B(J + 1)$. As mentioned before
(p. 14), the molecule will 'absorb' the photon and see if it is a suitable one. It will
take some time before it is discovered to be unsuitable, after which it will be rejected.

If it has not been discarded before the next collision, the energy (i.e. the photon)
may be transferred to the other molecule and hence it will look as though the first
molecule has absorbed the photon in the normal way.

The larger the interval between two collisions the better the control on the photon
frequency, the smaller the deviation from the 'true' value and the narrower the absorption band.

Let us take CO as an example

$$2B = 3\cdot84 \text{ cm}^{-1} \text{ and so}$$
$$\nu = 3 \times 10^{10} \times 3\cdot84 \approx 10^{11} \text{ Hz}$$

Thus the frequency of a suitable photon has to be 10^{11} Hz. The number of fully completed cycles between two collisions is

$$\frac{10^{11}}{5 \times 10^9} = 20$$

and so the inaccuracy in the photon frequency is $\frac{1}{20}$ or 5%. Hence collisions cause broadening of the 'absorption frequency,' or in other words, broadening of the rotational energy levels. As the frequency of a photon deviates from the exact frequency the probability of absorption decreases.

Since the collision frequency decreases at reduced pressure, bands will then become sharper. For instance the bandwidth for CO at 75 mm Hg at room temperature will be about $\pm0\cdot5\%$. Conversely, the bands will become broader at higher pressure as collision frequencies will be increased.

A similar consideration and calculation can be applied to the bandwidth of vibrational peaks. Suppose one has a peak at 3000 cm^{-1}. The frequency will thus be 10^{14} Hz. Provided the number of collisions per second is still about 5×10^9, the number of completed vibrations in between two collisions is

$$\frac{10^{14}}{5 \times 10^9} = 2 \times 10^4$$

So the inaccuracy in the band frequency will be $\pm5 \times 10^{-3}\%$ ($\pm0\cdot15$ cm^{-1}, negligibly small in comparison with other causes of band broadening (see next section).

Liquid phase. In the liquid phase, interaction between molecules is such that virtually no free rotation exists. Only the lowest E_{rot} levels play some role; the others will not be occupied. Besides, the lowest energy levels will be very broad due to the numerous collisions; they even might overlap each other. The few bands are no longer separated but form one diffuse relatively broad band. In the infrared this will give rise to a broadening of the vibrational bands, for as we already know, the rotational levels are superimposed on the vibrational levels. Naturally these levels themselves are also spread, due to the increased number of collisions, but this is insignificant in comparison with the broadening caused by the rotational levels.

Summarising, we may say that on changing from the gas phase into the liquid phase, the rotational fine structure disappears and the vibrational bands become less sharp. A frequency shift for all bands is very likely; it can be assumed to be due to the molecular interactions.

Solid phase. In the solid state a molecule is fixed in a crystal lattice. Rotation of the molecule as a whole is therefore prohibited; rotational levels no longer play a role. The result will be a sharpening of the vibrational bands compared to the liquid state. (This does not hold for molecules crystallising in an ionic lattice.)

Since organic molecules crystallise in general in unsymmetric lattices, a splitting up of bands may occur. Bands coinciding in the liquid state shift to different wavenumbers due to the fact that the corresponding modes are slightly different in the solid state structure. Even the same functional group (e.g. C=O) may show a split band as result of different orientation in the crystal lattice. It may therefore look as if two C=O functions are present, but a solution experiment can clear up any ambiguity.

Sometimes rather distinct bands appear which are not present in the solution or liquid phase. They may be attributed to vibrations of the lattice (partly or as a whole), as a result of, for example, the formation of associated molecules.

Finally it must be pointed out that some bands might show a considerable orientation effect (see p. 55) as result of instrumental polarisation, present in every spectrometer.

Triatomic molecules

A triatomic molecule may be either linear or bent, the number of fundamental vibrations being 4 and 3 respectively. Apart from the stretching vibration, a bending mode is possible for these systems. This can be demonstrated best by some examples (see Fig. 20). The movements of the atoms in the plane of the drawing relative to the

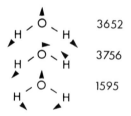

O - C - O	-	I	3312	H - C - N
O - C - O	2349	II	2089	H - C - N
O - C - O	667	III	712	H - C - N
+O - C - O+	667	IV	712	+H - C - N+

H O H   3652

H O H   3756

H O H   1595

FIG. 20. Modes of some triatomic molecules. The length of the arrows is not proportional to the real amplitude

centre of mass are indicated by arrows. An arrow represents only half a movement of course; in the second period the direction is opposite. Perpendicular modes are as usual indicated by + and − representing movements towards and away from the reader respectively.

Carbon dioxide. A linear symmetric molecule. Vibration I is called the totally symmetric stretching vibration and since there is no net alteration in dipole during the movement, this fundamental will be infrared inactive; no absorption band will appear. The second vibration (II) is called the asymmetric stretching vibration and since the dipole is altered, it will be infrared active. The next one (III) is a bending vibration, also active, while IV is the same vibration but in a plane perpendicular to that of III. This will not give rise to another absorption band, however, since both vibrations are fully identical. This phenomenon is called degeneracy.

Hence CO_2 will only produce two fundamental bands; stretching and bending ones at 2349 cm^{-1} and 667 cm^{-1} respectively.

Hydrogen cyanide. A linear non-symmetric molecule, the number of fundamentals still being 4. Number I (as well as II) is active now, due to the asymmetry of the molecule. Number I is called the CH stretching mode, while II is said to be the CN stretching mode. Here again as in the case of CO_2, III and IV are degenerate modes at 712 cm^{-1}.

Water. A symmetric bent molecule. Again I is called the symmetric stretching mode and II the asymmetric one. Vibration III is a symmetric bending mode. Movements out of plane will produce rotation only and can therefore be omitted. The spectrum of water will thus contain three fundamental absorptions (at 3756, 3652 and 1595 cm^{-1}).

More examples may be considered, but no matter what molecule is chosen, its vibrational modes belong to one of the three types represented here.

Combination bands

In addition to the fundamental and overtone bands, combination bands are found in the infrared region. The origin of such bands is rather simple. Consider a molecule with fundamental wavenumbers σ_1 and σ_2. It may also absorb energy quanta $hc\sigma_3$, where σ_3 is some linear combination of σ_1 and σ_2, i.e. $\sigma_3 = \sigma_1 + \sigma_2$, $\sigma_3 = 2\sigma_1 + \sigma_2$, $\sigma_3 = \sigma_1 - \sigma_2$ etc. Which combinations are allowed and which are not can be deduced from symmetry considerations. Like overtones, these combination bands arise only as a result of anharmonic oscillation, for interaction between two vibrations is then possible.

Degeneracy

As we have seen before, the number of normal vibrations for a molecule is $3N-6$ or, for a linear molecule, $3N-5$. This does not by any means denote the number of absorption bands that may be observed in the infrared. Whether a band *does* appear depends on several factors such as the selection rules, the intensity, the symmetry of the molecule and the infrared region under consideration.

If two or three vibrational modes are fully identical as result of symmetry, they are said to be degenerate; doubly degenerate as in the case of the symmetric bending vibration of CO_2 and many other molecules, or triply degenerate. The latter will only occur in molecules with a high symmetry, and are of no importance in common infrared work.

It may also happen that two or more vibrational modes of a molecule have the same(or nearly the same) difference in energy between two energy levels. The corresponding absorption bands, occurring at frequencies identical or very close to each other, will be indistinguishable. This phenomenon is called accidental degeneracy.

Summarising, we may say that degeneracy causes a decrease in the number of absorption bands that may be observed.

Isotope effect

When a particular atom in a molecule is replaced by its isotope there will neither be an appreciable change in internuclear distance nor in the binding force. There is only a change in mass, and hence in the reduced mass μ. Considering HCl as an example one can calculate the change in μ to be about 0.15% in going from $H^{35}Cl$ ($\mu = 0.9722$) to $H^{37}Cl$ ($\mu = 0.9737$). Since B is inversely proportional to μ, the rotational energy levels will be altered by 0.15% or about 0.03 cm^{-1}, obviously too small to be seen in the rotational spectrum.

As the vibrational frequency for HCl is proportional to $1/\sqrt{\mu}$, the difference between the energy levels for both isotopic molecules will be about 0.07%. For an absorption at 2886 cm^{-1} this means a difference of about 2 cm^{-1}, sufficiently large to be seen.

Since naturally-occurring HCl is a mixture of fairly large amounts of both chlorine isotopes, all absorption bands in which the vibrational energy of 2886 cm^{-1} is involved will appear as doublets, separation 2 cm^{-1}, provided spectrometer resolution is high enough.

A more drastic change in the energy levels for HCl – the vibrational as well as the rotational – is obtained when the H atom is replaced by its isotope D; the reduced mass increases from 0.97 to 1.89. As this is nearly a factor of 2, the rotational levels will be only half the distance apart (5 cm^{-1}), while the vibrational energy will be altered roughly by a factor of $1/\sqrt{2}$ or 0.7. Hence the vibrational band for DCl should be near 2000 cm^{-1}. (It is in fact observed at 2100 cm^{-1}.)

For an isolated C—H molecule, μ is 0.92. The reduced mass for the ^{13}C—H group is 0.93, while it is 1.71 for C—D. From this it can be concluded that the change in μ for a diatomic system will only be appreciable in the following cases: (a) a large difference in mass and (b) the lightest atom replaced by its isotope.

The only other possibility arises when both atoms have about the same mass. For instance, H_2 ($\mu = 2.00$), D_2 ($\mu = 1.00$), HD ($\mu = 1.50$), C—C ($\mu = 6.00$), C—^{13}C ($\mu = 6.24$) etc. It is apparent that the lighter the molecule the greater the isotope effect.

Since the B value for most molecules is small (less than a few cm^{-1}) the rotational energy levels will hardly be influenced by isotopic substitution, deuteration being an understandable exception of course in several cases.

The vibrational energies are even less influenced, as the change is proportional to $1/\sqrt{\mu}$, but because of the rather high ΔE_{vib} in comparison to ΔE_{rot}, the effect is far more pronounced. For instance, a change of 0.1% in μ will give rise to a shift of 1.5 cm^{-1} for an absorption band at 3000 cm^{-1}, visible under high resolution.

For a C—H stretching band the substitution of H by D changes μ roughly by a factor of 2. This will lead to a shift for this band from about 3000 cm^{-1} towards 2100 cm^{-1}, $\sigma_{\text{CH}}/\sigma_{\text{CD}}$ being proportional to $\sqrt{(\mu_{\text{CD}}/\mu_{\text{CH}})}$ or roughly $\sqrt{2} \approx 1.4$. In general the ratio $\sigma_{\text{XH}}/\sigma_{\text{XD}}$ will always be $\approx \sqrt{2}$ provided $m_X \gg m_H$. This effect can therefore be very useful for the identification of an X—H stretching band, on condition that that particular H atom can be replaced by deuterium without any other changes to the molecule.

POLYATOMIC MOLECULES

An increase in the number of atoms per molecule causes a further increase in the number of normal vibrations. True, the number will decrease as a result of symmetry (infrared inactive vibrations) and (accidental) degeneracy, but the spectrum would be hard to disentangle if it had a random distribution of all these bands. Fortunately it is found that compounds having a particular group in common do show absorption bands in the same region(s). For instance, all molecules with at least one CH_3 group absorb at 2900 cm^{-1}. An explanation is given below.

Functional groups

Consider different molecules in which a CH_3 group is present. If the vibrational transitions for a CH_3 group as such are completely different from any other transition in the rest of the molecule, the transitions or the modes of the CH_3 group will be independent; the group seems to be isolated. Remember, however, it is the molecule as a whole that is involved in each vibration and therefore, though small, a slight effect is to be found. The result is that a CH_3 group in a molecule absorbs in always the same narrow wavenumber regions (i.e. at about 2960 cm^{-1}; see also the Tables).

There are many more groups with rather unique vibrational energy transitions which behave like the CH_3 group; for example the groups OH, $C=O$, $C\equiv N$, NH_2, $C\equiv C$. They all give rise (if active) to certain absorption bands in well-localised regions in the spectrum.

Such groups are called functional groups as it is usually possible to conclude from an i.r. spectrum whether the group is present in a molecule or not.

Skeletal vibrations

Many other groups cannot be used as such, their accompanying absorption bands being very much influenced by the rest of the molecule. A good example of this is the single C—C band. If for instance two C—C bands are coupled – a very common situation in organic molecules – there will be an appreciable exchange of vibrational energy between the two groups.

This can be understood easily for, if free, both groups would absorb energy of the same frequency.

The result of the coupling will be that the places of the C—C absorption bands in a spectrum are strongly dependent on the adjacent bands or in other words the skeleton of the molecule. Vibrations of this type are often referred to as skeletal vibrations, and are highly specific for each molecule.

After this one may ask oneself what happens if two functional groups are attached to each other. Will they exchange energy strongly? Let us take two examples.

1. H_3C—CH_3. As the energy involved in the vibrations of a CH_3 group is quite different from that of a C—C group, the latter acts as an isolator; there will be no appreciable exchange of energy between both CH_3 groups.
2. $C=C$—$C=C$. Here too the C—C bond acts as an isolator, but less efficiently than that in the case of the ethane molecule. This is due to the fact that it has a slightly double bond character. Some interaction between both $C=C$ groups therefore takes place. Yet from practice it is known that in this case too, the $C=C$ group can still be called a functional group, as its absorption bands appear in the expected region.

Types of vibrations

The spectrum of a molecule will consist of several bands. As far as these bands can be unambiguously attributed to certain well-known vibrational modes of functional groups, these have names. We will mention the most important ones for they are

3

often used in the literature. Taking the methylene group as an example to start with, the movements and corresponding names are as indicated in Fig. 21. The notations underneath the names are common ones, though not the only ones possible.

Similar normal vibrations can be visualised for other groups, e.g. NH_2, NO_2, CH_3, etc. The same names are given to modes that are identical to those for the CH_2 group. For instance $\nu_{CH_3}^{s}$ is the symmetric stretching vibration where all three H atoms move up and down along the direction of the C—H bond. $\nu_{CH_3}^{as}$ is

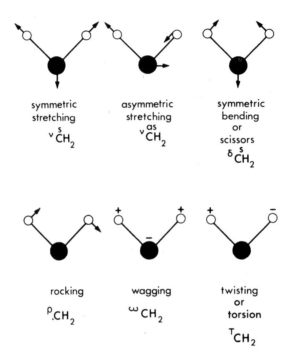

FIG. 21. Types of normal vibrations for a methylene group. ○ is the hydrogen, ● is the carbon atom

the asymmetric stretching vibration where two hydrogen atoms move upwards along the CH band while the other one moves downwards and vice versa. The others, $\delta_{CH_3}^{s}$, $\delta_{CH_3}^{as}$ and ρ_{CH_3} can be found in a similar way.

The modes of an OH group attached to a carbon skeleton can also be indicated by the above-mentioned names; ν_{OH}, the O—H stretching, ν_{C-O} the C—O stretching and δ_{OH}, the bending vibration.

A sub-division of some modes such as 'out of plane' and 'in plane' is frequently used, whereby the plane of the molecular skeleton is considered to be the plane.

LARGE MOLECULES

Finally, some remarks will be given about very large molecules such as steroids or even larger ones like proteins and polymers. In these cases, the number of atoms per molecule varies from about 50 to 1000 or more. Will the spectra not be too

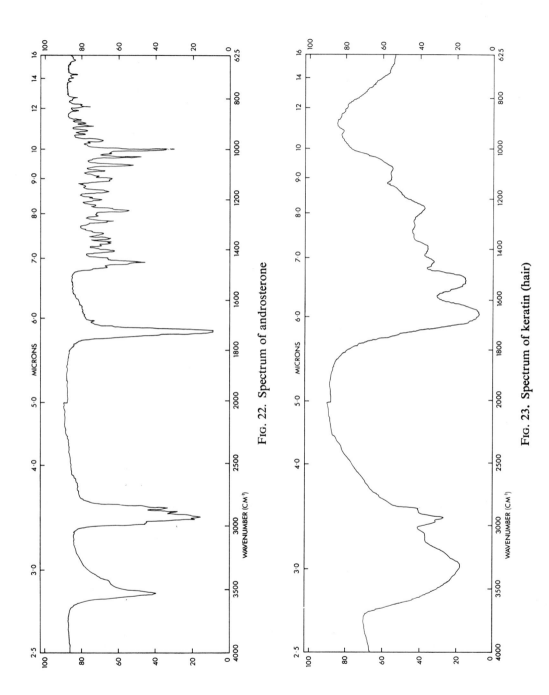

Fig. 22. Spectrum of androsterone

Fig. 23. Spectrum of keratin (hair)

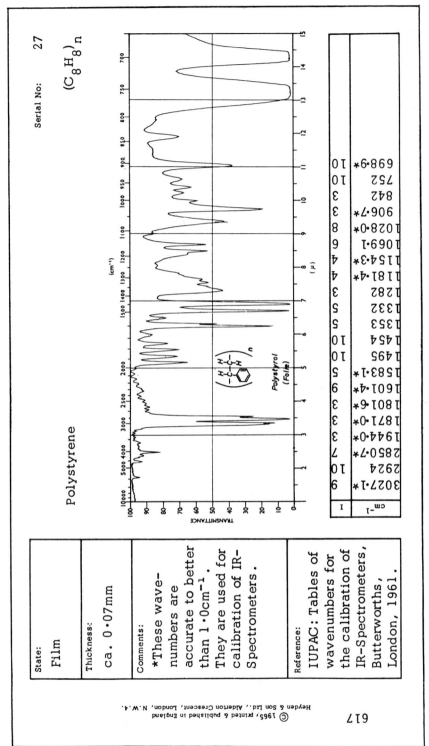

Fig. 24. Polystyrene spectrum, Mecke collection No. 22

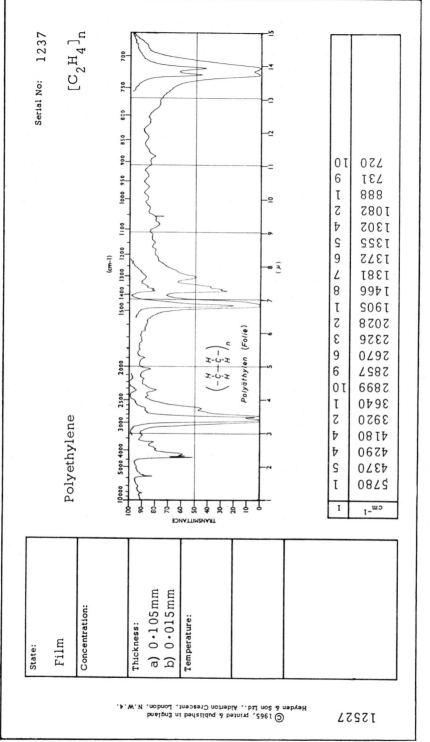

FIG. 25. Polyethylene spectrum, Mecke collection No. 1237

12527

complex for analysis, as the number of fundamental vibrations is already about 150 for the steroid? Besides, as steroids in general have no symmetry element, all these transitions are infrared active. Many vibrations, however, are to be ascribed to carbon–hydrogen modes and as such are observed in the same spectral region; they coincide or overlap and thus allow a simplification of the spectrum (Fig. 22). For such a huge molecule as a protein the situation is worse. There are so many groups with a slightly different character and the skeletal modes are so many that hardly any distinct band can be observed in such a spectrum. The many bands, though of different intensity, do overlap in practically all interesting regions. The result is a spectrum of extremely broad bands in which, however, some fine structure (peak maxima, shoulders) can be found (Fig. 23). Obviously, these spectra cannot be used for structure determination, but they may be of great help in comparing or recognising proteins.

Bearing in mind the spectra of proteins, one might be surprised to see that the spectrum of a polymer such as polystyrene (Fig. 24) is so simple, notwithstanding the large number of atoms per molecule. Here it is the simple composition of the polymer, in fact n times the same unit, that makes the spectrum look so simple

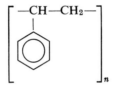

The spectrum can be considered to be an n-fold superposition of the obviously rather simple spectrum of the molecule between brackets. The carbon skeletal vibrations will contribute also, but as their intensity is usually small, they will not seriously interfere with the simplicity of the spectrum. It is not surprising either that the spectrum of polyethylene, $(CH_2)_n$, is even simpler than that of the molecule hexane, C_6H_{14}, as in the latter the CH_3 groups are extra contributors (Fig. 25).

BIBLIOGRAPHY

G. Herzberg, 'Infrared and Raman Spectra of Polyatomic Molecules,' *Molecular Spectra and Molecular Structure*, Van Nostrand, New Jersey, 1945.
K. W. F. Kohlrausch, *Der Smekal-Raman-Effekt, Ergänzungsband* 1931–1937, Springer Verlag, Berlin, 1938.

3

Spectrophotometers

INTRODUCTION

It is not the aim of this work to give in full detail the working and the construction of the normal type of i.r. spectrophotometers. The reader is referred to the literature as several books on this subject are available. The layout of a double beam spectrophotometer will be discussed using a block scheme:

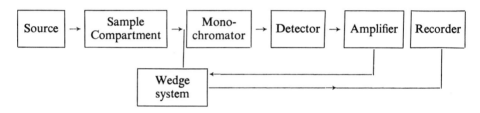

The optical diagram of a spectrometer can be found over the page (Fig. 26). The different parts of this spectrophotometer will be dealt with in subsequent sections below, except for the sample compartment, which is described in Chapter 4.

SOURCE

The most ideal light source would be one emitting constant energy over the whole infrared region. Unfortunately as yet such sources have not been developed and so we have to do the best we can with the following ones:

> Nernst filament (ZrO and some other rare earth oxides)
> Globar (Si–C)
> Ni–Cr wire
> Heated ceramic
> Mercury lamp

The main disadvantage of all these sources is the very unequal energy distribution in relation to the wavelength. The energy emission curve for 'black body radiation' is as given in Fig. 27. The curve is temperature dependent and can be calculated from Planck's formula

$$I_{\lambda,T} = \frac{2\pi hc^2 \lambda^{-5}}{\exp\left(\dfrac{hc}{kT\lambda}\right) - 1}$$

where I is the emitted energy, k is Boltzmann's constant and λ is the wavelength in nm. The wavelength with maximum energy output is found from

$$\lambda_{\max} T = \text{constant (Wien's law)}$$

whereas the total amount of emitted energy is proportional to T^4 (Stefan-Boltzmann law).

The curves for the above-mentioned sources can only at best be as good as a perfect Planck curve. As one wants a constant energy output from the monochromator a

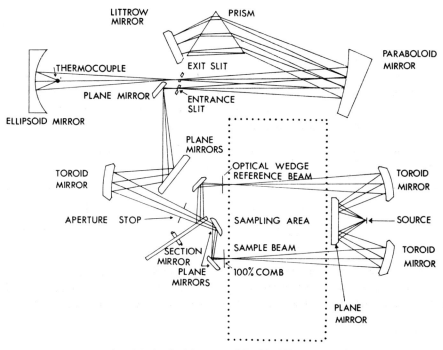

FIG. 26. Optical layout of Perkin Elmer 157

variable slit is necessary. The variation of the slit width has to be chosen in some way or another to be inversely proportional to the energy curve in order to be sure that a fairly constant amount of energy will reach the detector. To maintain a constant energy output the power supply for a source has to be stabilised. Fluctuations in voltage would cause fluctuations in the temperature of the source, and consequently changes in emitted energy.

In most cases the Nernst filament as well as the Globar have to be connected in series with ballast lamps to limit the current. Both sources have a negative temperature coefficient; i.e. the resistance lowers with increasing temperature. To prevent them from melting the current has to be limited.

MONOCHROMATOR

The system of slits, mirrors and prisms and/or grating, necessary for the dispersion of the radiation into separate wavelengths, is called the monochromator. The infrared beam passes from the entrance to the exit slit. The entrance slit can be viewed as the light source of the monochromator, its energy being fairly constant due to the variation of the slit width. Plane mirrors are used mainly in connection with the splitting and recombining of the two beams. They can also be used to make an instrument more compact. Non-planar mirrors (spherical, ellipsoidal, parabolic) are used either

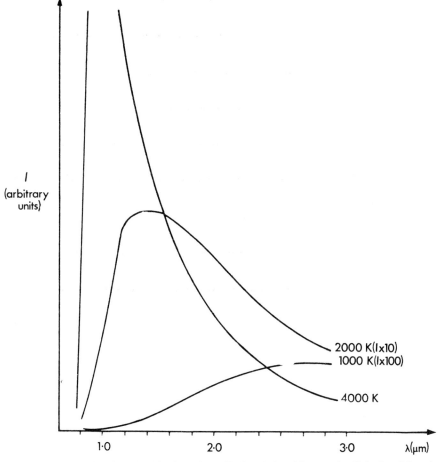

Fig. 27. Energy distribution for black body radiation (Planck's curve). Calculated for three temperatures from Planck's formula

to produce a parallel beam necessary to obtain satisfactory dispersion with a prism or a grating, or to form a reduced image of the source in the sample area or of the exit slit at the thermocouple. This slit is variable also, thus determining the 'monochromaticity' of the passing radiation. In fact both slits are normally identical.

Slit

As seen before there are at least two slits, an entrance and an exit slit. The width of both slits varies simultaneously with the variation of the wavelength. This can be done either mechanically (by means of a cam) or electrically (by means of a non-linear

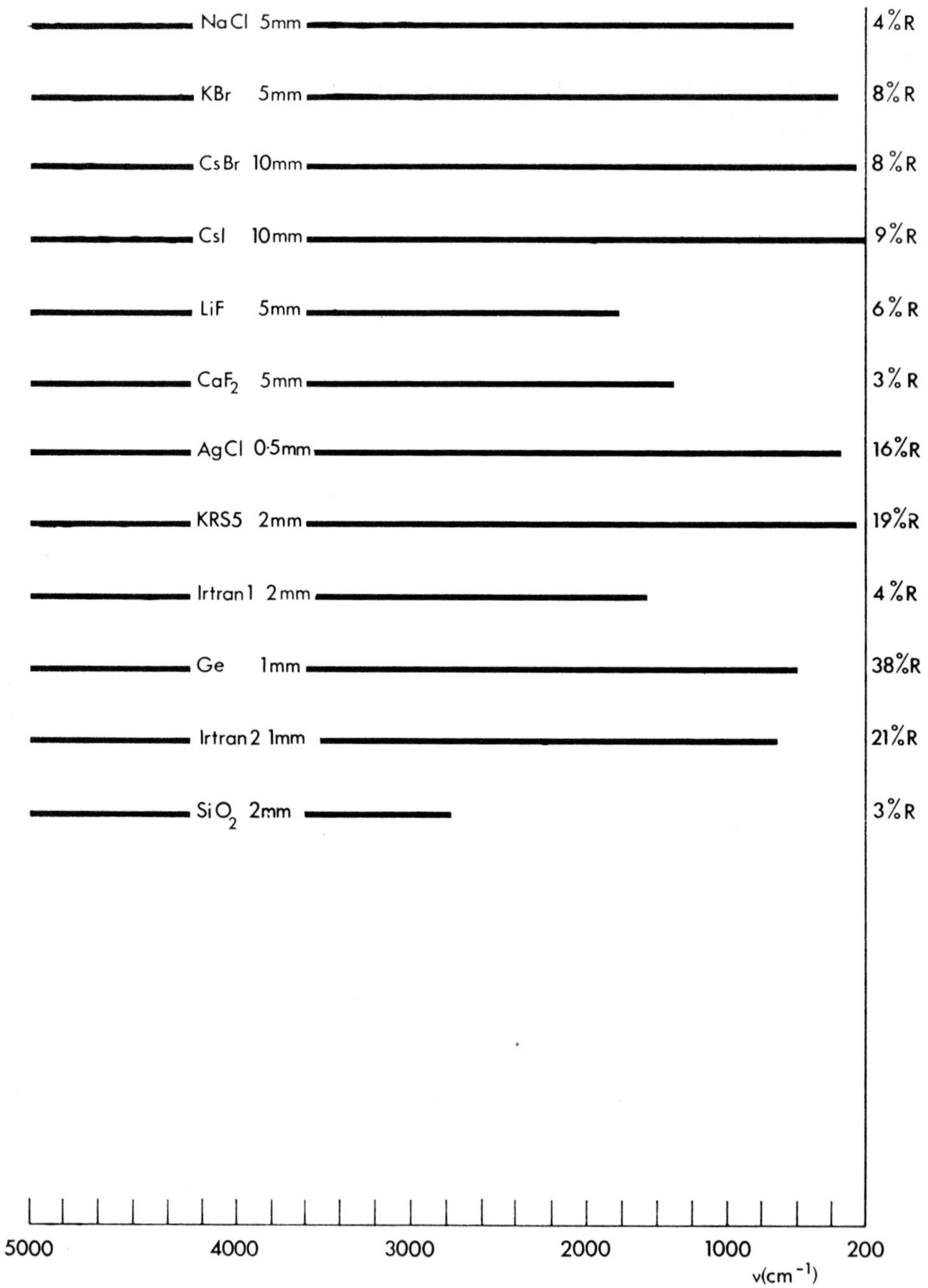

FIG. 28. Regions of $> 75\%T$ for several materials. The thickness is as indicated. The data in the right column refer to % reflection losses

potentiometer). Most spectrophotometers have more than one 'slit programme', i.e. at any wavelength the slit width can be changed by choosing another slit programme. On more expensive spectrometers the slit programme can even be altered continuously. Obviously a smaller slit requires an increase in amplification. An increase in slit width causes a decrease of the resolution. This of course is also true for normal increase of the slit width at lower wavenumbers, but for prism instruments this is more or less met by the increasing dispersion of the prism in that region.

Prism

Prisms for i.r. instruments are made from rock salt (NaCl), KBr, CsBr, LiF, CaF₂, KRS-5 etc. The prism material should be chosen in accordance with the resolution required. The dispersion of a prism depends on the change of the refractive index. The dispersion increases as the absorption band of the material is reached, though the n decreases rapidly. The transparency of the above-mentioned materials is shown in Fig. 28.

Some of the prisms are sensitive to moisture. Protection is obtained by elevating the temperature of the prism some 20 degrees above room temperature. The temperature has to be constant, as n, and thus the dispersion, changes with temperature.

Grating

Instead of the well-known prism as a medium of dispersion, a grating can be used for the same purpose. A grating is a mirror provided with numerous parallel lines or grooves. When a parallel light beam strikes such a grating, each line acts as a more or less infinitely small light source, emitting in all directions. Interference between the different rays from the different lines will occur. If the difference in path length between two successive rays of the same wavelength equals k times the wavelength ($k = 1, 2, 3 \ldots$ an integer) the intensity will be enhanced, whereas for $(k + \frac{1}{2})$ times, extinction will be the result (see Fig. 29).

This mathematically expressed as

$$d[\sin \beta + \sin (\alpha + \beta)] = k \cdot \lambda$$

where d is the distance between two parallel grooves, i.e. the grating constant, λ is the wavelength of the rays looked at, and β and $(\beta + \alpha)$ the angles between the incident and diffracted beam and the normal to the plane of the grating.

For a given spectrometer the small angle α is constant since the entrance and exit slit are in a fixed position. Suppose now $k = 1$ (first order) then, since d is constant (inversely proportional to the number of lines per mm), the diffracted rays will form a spectrum because, for increasing values of β, and thus $\sin \beta$, the wavelength of the rays will also increase. The spectrum will resemble one formed by a prism.

A grating will, however, produce more spectra, e.g. a second one exists for $k = 2$ (second-order spectrum) etc. Unfortunately these spectra do show overlap as can be easily understood from the formula: λ in the first-order spectrum is overlapped by $\frac{1}{2}\lambda$ of the second order, by $\frac{1}{3}\lambda$ of the third order etc. A small prism or filter is therefore needed to separate one order from another.

The intensity of the diffracted rays varies with both the angle β and the order. The larger β (and thus the wavelength) the smaller the intensity, and also the higher

the order the lesser the intensity. In general the first order is used as this gives the widest possible range of wavelengths together with less stringent filtering requirements.

From the formula one obtains easily that for $\beta \to 90°$ the wavelength of the diffracted beam will be maximum, $2d = k\lambda$ or $\lambda = 2d$ for $k = 1$ and $\alpha \ll \beta$. Such large β values cannot be used for reason of lack of light yield.

To improve the light yield in the first order at a distinct wavenumber, echelette type gratings are produced. The form of such a grating is drawn schematically in

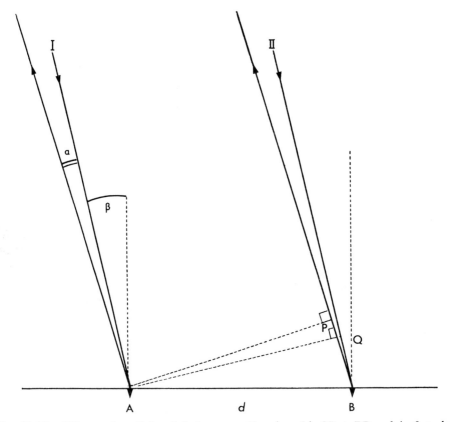

FIG. 29. The difference in path length between ray II and ray I is QB + BP or $d \sin \beta + d \sin (\alpha + \beta)$ respectively

Fig. 30. The angle ϕ is called the blaze angle. It determines at what wavelength the light intensity will reach a maximum ($\beta \approx \phi$). The grating is said to be blazed at that wavelength or wavenumber.

DETECTOR

The dispersed light passing the exit slit is focused with the help of a concave mirror onto a detector. In present-day instruments thermal detectors are used, i.e. the radiation is converted into thermal energy and the change in temperature is detected by either a thermocouple or a Golay cell. Several other detectors do exist but are still rarely in use at present (Bolometer, photo-tube etc.).

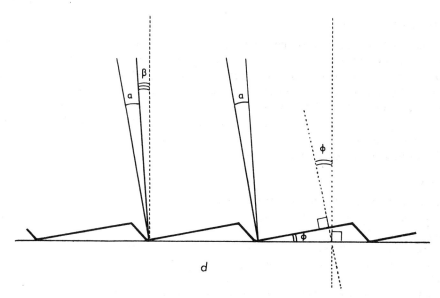

FIG. 30. Echelette type grating with blaze angle ϕ

Thermocouple

Though specially constructed for its use in i.r. spectrometers, the action of the thermocouple is quite normal. A rise in temperature causes an increase in the electrical potential. Since the sample and the reference beam are focused onto the detector in turn the potential of the thermocouple will vary with time as indicated in Fig. 31, provided the sample is absorbing some energy. The period of this variation is equal to the time required for the chopper to complete one cycle.

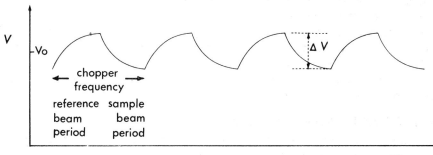

FIG. 31. Output signal thermocouple. (Reference and sample beam unequal)

The output of the thermocouple is thus a d.c. potential V_0 plus an a.c. potential of amplitude ΔV and frequency equal to the chopper frequency.

The thermocouple is mounted in a small evacuated housing provided with an i.r. transparent window. The effect of the vacuum is to eliminate the loss of heat from the target to the air, resulting in about a four-fold increase in the temperature rise of the target in response to a fixed amount of radiation.

Golay cell

A Golay detector is a pneumatic detector; i.e. radiation is converted into heat, causing the expansion of a gas. The expanding gas alters the position of a mirror that forms part of an optical pathway ending with a photo-cell. The higher the temperature, the greater the expansion, the greater the distortion of the optical Golay system and the smaller the amount of light falling onto the photo-cell.

The n.e.p. (see below) of this detector equals that of the thermocouple or may even be somewhat better. The Golay cell is more fragile than a thermocouple, however, and thus its lifetime is liable to be shorter.

Noise

The 'noise' of a spectrophotometer can be defined as the average random movement of the recorder pen, peak to peak, when both beams are unobstructed. It can be measured in % transmittance. It is primarily due to the 'Johnson noise' of the detector – the small, spontaneous variations of potential which are an inherent result of its electrical resistance. By comparison, the noise produced by the amplifier and mechanical system can, in general, be neglected. (The equivalent to the Johnson noise, in the case of the Golay cell, is the thermal motion of the expansion gas.)

A good measure of the performance of a detector is its 'noise equivalent power' (n.e.p.). This is the power (in watts) of the radiation incident on the detector necessary to produce a signal equal to the noise under certain standard conditions. A typical thermocouple has a n.e.p. of about 10^{-10} watt and a sensitivity of about 10 volts per watt. (Note: the lower the n.e.p. the better the performance.)

Chopper

Though the word chopper is adopted from single beam instruments (the radiation beam is interrupted (chopped) for half a period by a rotating disk), it is also used in double beam instruments for the mirror system that causes the reference and sample beam to fall alternately on to the detector.

The chopper frequency has an upper limit determined by the response time of the detector. For a thermal detector, such as a thermocouple or Golay cell, this limit is typically between 10 and 20 Hz. In addition, to minimise interference from a.c. mains, the chopper frequency should not be near $1/n$ times (n is an integer) the mains frequency or else should be *synchronously* $\frac{1}{2}n$ times this frequency.

AMPLIFIER AND WEDGE SYSTEM

Most infrared spectrophotometers operate on the 'optical null' principle; that is, any attenuation of the sample beam by the sample is balanced by a corresponding attenuation of the reference beam by the 'wedge' (variable optical attenuator) so that both beams remain equal in intensity. When the two beams become unequal, due to a change in sample absorption, the detector produces an a.c. signal proportional to the difference. This signal is amplified and fed to a special electric motor – a servo-motor – which turns clockwise or anti-clockwise depending on whether the sample beam is more or less intense than the reference beam. The rotation of the servo-motor causes the wedge to move into or out of the reference beam until both beams are equal again. The a.c. detector signal then becomes zero and the servo-motor stops. Even with complete unbalance of the beams the a.c. signal from a thermocouple is very small (about $0.1\ \mu$V). This cannot be sent very far before amplification is necessary, and this is performed in the pre-amplifier, after which it is sent to the main amplifier, which is tuned to the chopper frequency.

The form of the wedge is such that its position in the reference beam is linearly proportional to the transmitted energy. Thus the recording of the position of the wedge as a function of frequency (wavenumber) amounts to the same as recording a spectrum.

The coupling of recorder pen and wedge system may be either mechanical (built-in recorders only) or electrical, by means of a potentiometer coupled mechanically to the wedge.

SCANNING CONDITIONS

Introduction

On each spectrometer, no matter how simple it is, one will find several adjustable and/or preset controls to operate the instrument. A manual will be helpful in setting them in the correct position for routine work, but one may have one's own wishes and ideas, and must therefore grow accustomed to the use of these controls in order to get the desired result.

The infrared spectrum of a compound may present a great deal of information about the substance provided the instrument is used carefully. The spectrum is unique and specific; each absorption band is unequivocally characterised by its intensity and its wavenumber, the former related to the number of molecules in the beam, the latter somewhat dependent on the sampling technique. To get the best spectrum we can, the spectrometer has to be adjusted to optimum condition. Many factors influencing the overall result cannot be altered by the operator (e.g. optical path). Some factors have to be checked fairly regularly (e.g. wavenumber calibration, balance, zero and 100% line etc.), while others are to be chosen before the spectrum is scanned (e.g. scanning speed, slit programme, gain etc.). Several of these factors or instrumental parameters will be examined briefly in the following sections.

Scanning speed

The scanning speed is limited by the response time of the comb/servo system, a system that indicates the absorption by the sample. If the scanning speed is too high, the servo system will not be able to follow the absorption pattern caused by the sample. The absorption bands, especially those with a steep slope and of great intensity, will be deformed. Neither the intensity nor the wavenumber of these bands will be correctly recorded (Fig. 32).

The optimum scanning speed is the maximum speed at which any decrease does not change the shape and the exact place of the absorption bands. Since 'time is money', it should not be chosen smaller than necessary.

Some spectrometers are provided with automatic speed suppression (a.s.s.). The scanning speed can then be set rather high, for as soon as an absorption band appears, the a.s.s. causes a decrease in the speed during the recording of that peak.

Correct settings of the speed and the a.s.s. can be found by trial and error. A speed reduction or a higher a.s.s. must not alter with respect to the shape and the position of the bands.

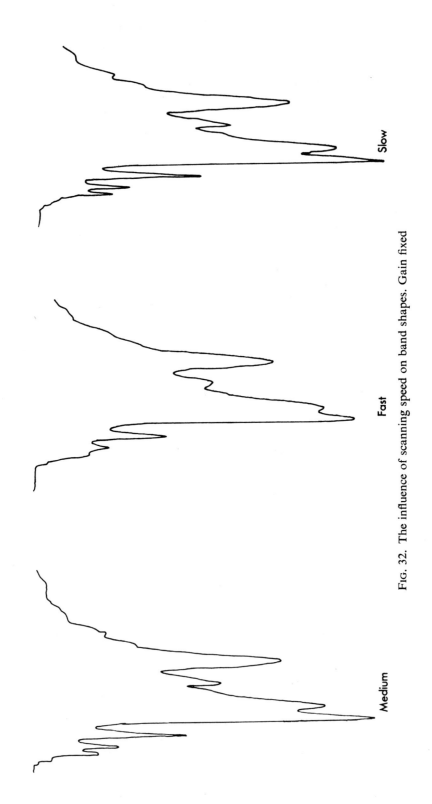

FIG. 32. The influence of scanning speed on band shapes. Gain fixed

Slit programme and resolution

As we saw before, the slit width can be chosen between certain values. The narrower the slit the better the monochromaticity of the radiation passing the exit slit. The dispersion of the radiation itself depends on the construction of the monochromator (mirrors, focal length grating and/or prism etc.). For a monochromator the resolution or resolving power (R) is defined as $\nu/d\nu$, where $d\nu$ is the difference in wavenumber between two wavenumbers that can just be distinguished by the spectrometer. The ultimate or theoretical R is equal to $b\,dn/d\lambda$ for a prism and to $k \cdot m$ for a grating, where b is the base length of the prism, $dn/d\lambda$ the dispersion, k the order of the grating and m the total number of lines. Obviously for a prism the limiting R varies with wavelength but it does not for a given grating in a given order. In practice R varies more or

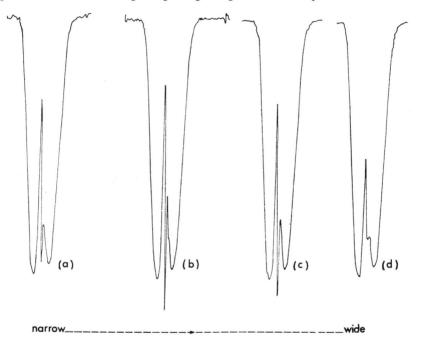

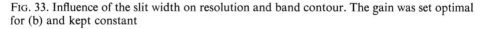

narrow_ _wide

FIG. 33. Influence of the slit width on resolution and band contour. The gain was set optimal for (b) and kept constant

less just as the slit width does. At a certain wavenumber the resolution can be changed by changing the slit programme from narrow to wide or vice versa. We have to choose the slit programme such that the instrument will produce the correct absorption spectrum. If the slit is too broad, resolution will be too low and the radiation will not be sufficiently monochromatic; bands or peaks lying close together will be deformed, and an average pattern will be the result (see Fig. 33). As we can see, bands become broader, intensities smaller, and details disappear. If the slit width is too small, however, bands may not show their true shape either unless the gain is appropriately increased (provided the limit of the gain is not reached and/or the resulting noise is still acceptable).

In conclusion one may say that the choice of the slit programme is important.

Gain

The amplification of the signal, necessary to get the comb/servo system, and thus the recorder pen in motion, is determined by the setting of the gain control. Some resistance has to be overcome to get the system moving, i.e. without a minimum amplification the pen will be 'dead'. This minimum is related to the amount of radiation (and thus to the slit width) that reaches the detector. On the other hand there is a maximum amplification value also depending on the slit width. The higher the gain the faster the servo system moves to a correct off-balance signal. For too high a gain setting the servo system will pass the correct position, thus producing an opposite signal. It will therefore stop and return but it will pass the correct position and so on. This continuous movement of the servo or pen is called 'hunting', and the passing of

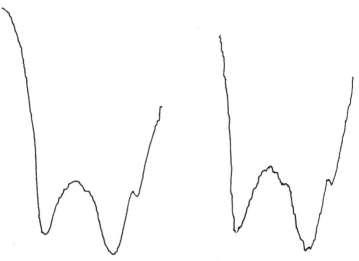

Fig. 34. Influence of noise on the peak maximum

the correct position is called overshoot. If the overshoot is not too severe the pen will eventually stop by itself at the correct position (damped oscillation).

Another parameter connected with an increase in gain is the noise caused by the detector and electronics. The higher the gain the greater the noise and vice versa. For accurate wavenumber measurements a high noise level is undesirable as an absorption band distorted with noise is hard to record exactly (see Fig. 34).

Luft formula

In 1947 Luft* found the following empirical formula

$$R^4(S/N)^2t = \text{constant}$$

where R is the resolution of the instrument, strongly related to the earlier-mentioned slit width or slit programme, S/N is the signal-to-noise ratio related to the gain in connection with the noise (N), and t is the scanning speed.

The formula is correct for most values of R barring very high resolution where the fifth power has to be used.

Let us illustrate the formula by some examples. Say an increase in the resolution of

* K. F. Luft, *Angew. Chem.* **B19**, 2 (1947).

a factor of 2 is required. This can be achieved at the cost of either a decrease in the speed by a factor of 16 or an increase in the noise – a worsening of the signal-to-noise ratio – by a factor of 4. Conversely a two-fold decrease of the resolution enables us to reduce the scanning time to $\frac{1}{16}$ of the original one.

This reasoning is only applicable if the scanning conditions are chosen or set optimally. For example, a two-fold reduction of the resolution *may* be followed by a 16-fold reduction in the speed, but it is not *necessarily* carried out; no coupling between the parameters exists beyond human intervention. Yet a smaller reduction is useless, while a greater one is inadmissible.

The question arises as to how one should adjust the three parameters correctly. If one combination is known the others can be deduced from Luft's formula. To find one of the possible combinations, one may proceed as follows:

1. Set the slit programme to normal; i.e. set the slit programme control to position N or to the recommended position; if there is an energy control instead set this one to normal or recommended energy.
2. Choose the scanning time. The entire spectrum should be run in about 30 minutes.
3. Adjust the gain control for a suitable signal amplification. This can be done in two ways. Both tests should be carried out in a region where there is no atmospheric absorption, e.g. at 2000 cm^{-1}.

Overshoot method. Adjust the 100% control such that the pen servo system indicates 90% transmittance. Bring an object into the sample beam until the transmittance is decreased to 80%. Withdraw the object at once and adjust the gain so that the pen shows an overshoot of 1%.

Dead spot test. Place an object (wire grid) in the sample beam and/or adjust the 100% control so the pen indicates about 50% transmittance. Block one beam and withdraw slowly. Repeat with the other beam. The pen should indicate the same transmittance in both cases (within the limits of the noise of course). If not, the gain should be re-adjusted. After this proceed as follows:

1. Adjust the balance control. Block both beams and see that no drift or a very small upscale drift is present. Otherwise readjust.
2. Run a test spectrum, for instance polystyrene. In general a sample showing broad and sharp bands as well as shoulders should be used. It must also possess some well-defined peaks (wavenumber accurately known).
3. Next re-scan the spectrum at lower speed, e.g. half the original one. Compare the absorption ratios of a pair of bands in both spectra; preferably a sharp and a broad band lying close to each other. If these are identical the original setting was acceptable and perhaps even a faster speed is possible. This can be checked in the same way. If both spectra are not identical the scanning speed has to be decreased further. This procedure should be repeated until both spectra are essentially identical.

If the instrument is supplied with other controls related to the scanning conditions, such as automatic speed suppression, a similar procedure is to be followed (see Potts and Lee Smith*).

* W. J. Potts and A. Lee Smith, *Applied Optics* **6**, 257 (1967).

The settings once found can be changed later in accordance with Luft's formula to other combinations for special purposes. A final check as mentioned above is required, however.

High resolution

Spectra can be obtained by narrowing the slit to its smallest value at a certain wave-number. Say the slit width is decreased by a factor p, then the energy is decreased by p^2 (two slits!) and so the gain has to be increased by p^2 to compensate for the energy losses. The noise will be increased as well by the same factor p^2. In the case of instruments equipped with a time-constant control, the noise can be reduced using this control, provided the scanning speed is adequately reduced. The final choice will depend on the capabilities of the spectrometer and on the personal preference of the operator (e.g. p in the gain and thus still p^2 in the speed). The final settings are to be controlled by running a test spectrum.

Limited energy spectra

If for any reason the background line shows a transmittance of about 90%, the ideal situation (background = 100% line) can be restored by placing into the reference beam a 10% absorbing accessory. It is found from practice that the background may have any value between zero and 100% transmittance and a variable absorbing accessory (reference beam attenuator) is useful. The restoring of the normal situation for double beam spectrometers by using an attenuator cannot, however, be done without repercussions. Suppose that a micropellet removes 60% of the available light. Now whether an attenuator is used or not, only 40% of the light is left and useful for absorption and detection purposes. Or, in other words, the available energy is decreased by a factor of 0·4. In order to get back to the original spectrometer settings, the gain or the slit or the speed are to be changed. If the light losses are very serious (ultra micro work, beam condensers), e.g. up to 90% or even higher, all three conditions have to be altered:

1. The speed as slow as possible.
2. The slit programme as wide as possible.
3. The gain as high as tolerable/useful with respect to the noise.

Even then the spectrometer settings may not be optimum and a slightly distorted spectrum may be the result. The usefulness of the spectra obtained in that way have to be controlled by comparing with spectra from the same compound run under normal (routine) conditions, but a somewhat distorted spectrum is often better than none!

Stray light

Light of a wavenumber $\neq \sigma$ reaching the detector simultaneously with light of wavenumber σ is called stray light. Provided the stray light is chopped (has passed the chopper) it will be amplified in the same way as the normal light beam signal. A false or partly untrue signal will be the result. Stray light is only permissible as long as its energy is very small; less than 0·5% in comparison to the energy of the real radiation. Stray light is the result of unwanted reflections and marginal rays just missing the optics. To minimise the stray light a monochromator is divided in compartments separated by black painted screens and if necessary provided with small holes.

Nevertheless some stray light does exist. The wider the slits the more stray light is to be expected. As we saw before the energy distribution over the complete wavenumber region is anything but flat. An energy ratio of 200:1 for 4000 and 600 cm^{-1} is quite normal. Hence it is clear that the stray light problem will be most important in the low wavenumber region: the slits are widest there and about 0·1% stray light of 4000 cm^{-1} causes a 20% error.

For qualitative analysis a total stray light percentage up to 2% is admissible. A straightforward reduction is possible by using special filters and/or double monochromators. More accurate quantitative measurements can be then carried out.

Automatic gain and slit control

In double beam instruments the energy falling onto the detector is not the same for all wavenumbers, despite the moving slits. For instance, in the regions where atmospheric absorptions are present (CO_2, H_2O) the remaining energy, if any, is low. The same holds for the regions where solvents absorb when the compensating technique is followed.

As the signal from the detector is linearly amplified, the final signal, the one supplied to the servo system, will vary greatly with the wavenumber. Having set the gain control to an optimum value at a certain wavelength (100% transmittance!) it will be wrong at all wavenumbers where atmospheric or solvent absorptions take place. The pen servo system will be slow to respond there, and a false spectrum may be the result. This problem can be overcome by changing either the slit width or the gain in those regions. Some spectrophotometers are already equipped with such devices, called automatic gain control (a.g.c.) or automatic slit control (a.s.c.) respectively.

In the case of a.s.c. the energy of the reference beam is kept constant by altering the slit width when passing an atmospheric or solvent band. Unfortunately resolution will be altered simultaneously. In the case of a.g.c. the gain is usually adjusted for minimum energy (at the wavenumber where atmospheric or solvent absorption is highest). For higher energies during scanning the a.g.c. system will then reduce the gain. Of course such a system produces variable noise in the spectrum; low noise at high energies and vice versa. Nevertheless it may be of great help when 'difficult' solvents are to be used.

Scale expansion

Expansion of the abscissa is frequently referred to as scale expansion. Though strictly speaking this may be permissible, the expression 'scale expansion' is to be restricted to an expansion of the ordinate. Few instruments are equipped with such a built-in unit, though some can be supplied with such an accessory.

This feature can be used when only a very small amount of sample in too low a concentration is available. The resulting spectrum will normally be too 'thin', i.e. the absorption pattern will be recorded between 100 and 90% transmittance. A 5 times scale expansion will now extend this region to 50%, while for 10 times expansion 0% transmittance will be reached again.

At first glance this may look promising and simple to do, the expansion being carried out either electrically or mechanically. At least two disadvantages are readily found, however:

(a) The noise is expanded as well by the same factor. For instance, from 0.2% to 2% for 10 times expansion.

(b) When the background does not coincide with the 100% line (a normal situation), the difference between both lines is expanded as well.

This is disastrous for 'background lines' showing a slope from 50% at 5000 cm^{-1} to 100% at 600 cm^{-1}, or more. The maximum expansion factor is now 2, as can be easily verified. Since the situation is usually worse for solids (due to reflection losses etc. especially at high wavenumbers) the expansion factor is limited for that reason. Yet scale expansion by moderate factors can be useful, for time is saved if the concentration or the cell thickness is too low.

Calibration

Calibration of an instrument should not only be done when installing the apparatus but also fairly regularly thereafter. There are two important things to be checked very carefully: the accuracy of the transmittance scale and the accuracy of the wavenumber (frequency or wavelength) scale. Other tests can be carried out, but these may be different for all instruments.

Transmittance scale. This scale can be easily checked by using in the sample beam fast-rotating sectors of known transmittances. A high speed of rotation is necessary (up to 2000 r.p.m.) to avoid interference between the rotating sector and the chopper motor frequency.

If the indicated transmittance is false, this can be restored by a re-adjustment of the comb, source or thermocouple though this latter can be a difficult and laborious job. For errors beyond the tolerated limit the instalment of a new comb is required. A periodical check is desirable, say about once a month when intensity measurements form an important part of the work and provided there is no significant change during this period.

For qualitative measurements, such as structure elucidation, the periodical check can be less frequent, say once a year.

Wavenumber/wavelength scale. The wavenumber or wavelength check has to be done at least once a week and even once or twice a day if very accurate measurements are to be performed. For this purpose the rotational bands of gases such as CO, CH_4, H_2O and NH_3 are used. The gas is let into a gas cell (path length 10 cm) and the spectrum is recorded under certain scanning conditions. Very accurate data for these rotational bands may be found in the literature (see Appendix A).

When the utmost accuracy is not essential, other more convenient materials may be used, such as polystyrene and indene. False wavenumber indication (i.e. error exceeding the tolerance) can be eliminated by re-adjustment. The prism-littrow system or the grating and the wavenumber scale indicator or marker should be disconnected. The exact position should be found by trial and error and then the systems should be reconnected.

The deviation can be either systematic (all wavenumbers being somewhat too high or too low) or variable. In the former case the error can be overcome by the above mentioned re-adjustment procedure.

In the latter case some misalignment in the optical path of the instrument must be present. This can be corrected best by a service engineer. It is not always necessary to correct small deviations (a few wavenumbers or a hundredth of a micron for example), as one can compensate for them provided the correction factor is known. If the spectrometer uses pre-printed sheets of chart paper, it is even simpler to insert them so that the error (if a systematic one!) is corrected directly.

Recording paper. The recording paper is a source of misunderstanding. One can distinguish two different types of recording: on loose pre-printed charts or on rolls of unprinted paper.

In the former case the size of a spectrum is completely fixed. For most spectrometers, however, these sizes are too small in view of the accuracy with which the wavenumbers can be determined. The exact size can be easily calculated as follows.

Suppose the average accuracy to be 4 cm^{-1}. If a spectrum is produced from 4000 to 600 cm^{-1} about 3400/4 clearly distinguishable positions are required to define the total amount of information. Furthermore, supposing that each position requires 1 mm, which seems to be fairly reasonable, the length of the spectrum should be about 850 mm.

Since the accuracy of the transmittance is not better than 1%, the number of positions is 100 and the vertical dimension should be therefore at least 100 mm. It is convenient however, in view of the overall image of a spectrum, to use a larger vertical dimension. From the above calculations, though rather rough, it is obvious that many pre-printed charts are too small.

Spectrometers recording on unprinted, blank paper in general can produce spectra of good dimensions. The only disadvantage, but a very important one in service work, is the impossibility of reading the wavenumber of a band straight off. Instead each band has to be determined by making use of some type of marking mechanism, which in its turn has to be calibrated on polystyrene or gases. With this type of spectrometer expansion is easily done.

With increasing automation flow chart recording on pre-printed paper is also possible now. This enables us to use either pre-printed charts with the advantage of an easy and exact read-off, provided the dimensions are right, or unprinted sheets with the possibility of expansion of the abscissa. A marking accessory is usually available on request.

Paper shrink and stretch. Due to changes in humidity or temperature chart paper is subject to shrink and stretch. For 'ready-to-hand' spectra this is unimportant, as well as for unprinted sheets or notes. The situation is different for pre-printed charts, since stretching or shrinking of these charts before the recording of a spectrum may result in false registration. A check on this point in advance is advisable.

BIBLIOGRAPHY

A. E. Martin, *Infrared Instrumentation and Techniques*, Elsevier, Amsterdam, 1966.
For calibration, see 'Tables of Wavenumbers for the Calibration of Infrared Spectrometers,' *Pure Appl. Chem.* **1,** 4 (1967).

4
Sampling

The appearance of a spectrum depends to a large extent on the preparation technique that is followed. Gas, liquid and solid spectra differ a great deal, as one knows from theory. Different spectra can, however, also be expected as a result of the preparation technique, the pre-handling of the compound and last, but not least, personal experience. What is really required is a spectrum containing as much information as possible. No general rule or rules can be outlined to achieve this, but a survey of the most common techniques will be of some help.

The different techniques are discussed briefly in the following sections, though no procedures are given. More details are to be found in the literature and/or handbooks.

GAS

A gas is most simple to handle; no mixing with other materials is required and thus it is easy to remove. Of the three phases it is however the one with the lowest density and consequently a rather large volume is required to obtain a reasonable spectrum.

A *gas cell* usually consists of an evacuable space supplied with two infrared transparent windows and at least one valve. Common dimensions are a path length of 10 cm and a volume of about 170 ml. It is evacuated first and then a gas is let in. The amount of gas in a cell can be diminished by lowering the pressure. The main disadvantage is the rather high 'dead' volume; only a small part of the gas is irradiated and thus useful for measurements. This can be overcome by constructing a cell that is tightly fitting round the radiation beam. Such cells are called *minimum volume cells*. They are commercially available with a path length of 7·5 cm and a volume of 25 ml. They are as easy to handle as the normal gas cell. Yet the volume in relation to the path length is a serious drawback for this type of cell.

A great step forward is the *multi-pass cell*. In these cells a considerable increase in path length is obtained with the help of a couple of inside mirrors. The radiation is reflected that way several times without the volume being altered. Path lengths of 1 m or even about 20 m are then possible. As result of the many reflections, however, the transmittance is decreased to 30% or less (due to stray light and reflection losses) and so the signal-to-noise ratio is influenced unfavourably. Other disadvantages are:

(a) The double-beam character of the spectrophotometer is completely distorted when a multi-pass cell is used. Water and carbon dioxide bands will no longer be compensated for.

(b) The cell is difficult to install, and minor changes in the adjustment cause a dramatic decrease in the transmittance.

(c) Although the efficiency for a multi-pass cell with respect to the gas volume is much higher than for an ordinary cell, it has been found to be necessary to increase the volume for the construction of cells with large path lengths. Therefore the absolute amount of gas required to fill a cell is increased.

A spectrum of a gaseous compound may also be obtained in solution. This method can be of importance if only small quantities of a gas are available in an excess of others, such as air. By dissolving the gas in a suitable solvent it is also concentrated and N_2 and O_2 do not interfere. The spectra of course are no longer gas spectra, but have great similarity to liquid spectra. Obviously the absorption of the solvent is a serious drawback of this technique. Furthermore, this method is more laborious than the preceding ones.

LIQUID

The liquid compound as such can be transferred into a *liquid cell* of fixed path length, 10 to 25 μm, by means of a syringe or, even simpler, by adding a few drops to one of the supplying channels. The disadvantages of this method are numerous.

(a) Such thin cells are hard to clean in many cases.
(b) The cells have rather high dead volumes, viz. the supplying channels and the non-irradiated part.
(c) The path length may be too large in some cases and too small in others.
(d) The recovery of the substance is difficult.

The amount of liquid can be minimised by making use of *minimum volume cells, microcells* or *cavity cells*. These have no longer an important dead volume, but they are still difficult to clean, especially when viscous liquids are involved.

The easiest way to handle a liquid is to introduce a few drops between windows. If the path length is important a spacer can be used, but it may be left out equally well. In general, the thickness can be adjusted by altering the pressure by which both windows are pressed together. The window technique cannot be used with too volatile compounds nor with unstable ones. Contact with air, though limited to the outer area, can be ruinous. This method is simple, however, and cleaning can easily be done afterwards. For quantitative measurements liquid cells with larger path length are to be preferred. The compound has to be dissolved; the spectrum is that of a solution. For cells of 0·1 and 0·5 mm a concentration of 5 and 1% w/v will usually do. Cleaning of such cells is no longer a problem. A limiting factor of course is the solvent's own absorption, despite compensation techniques. Therefore at least two different solvents, e.g. CCl_4 and CS_2, are required to cover the normal i.r. range, 2–16 μm.

Volatile or fairly volatile liquids may be transferred to a gas cell. This can be more useful than liquid cells as gas bubbles form easily and are difficult to remove. If the vapour pressure is too low to obtain reasonable spectra an increase in temperature might be helpful.

SOLID

A solid compound can be dissolved in a suitable solvent, after which the solution is transferred into a liquid cell. The disadvantages are as mentioned above for liquids.

Moreover the choice of an acceptable solvent, taking into account the solubility of the compound and the absorption of the solvent, will be very difficult or even impossible.

The *mull technique* is another common way to obtain the spectrum of a solid. Finely powdered material is homogeneously mixed with an inert, fairly transparent liquid such as paraffin oil. This is done to overcome the disastrous amount of scattering of light that would occur if the powder was brought into the sample beam between two NaCl or KBr windows. Disadvantages are obviously the absorption of the mulling agent, despite compensation techniques, and inhomogeneity due to gravitation. The technique is to be recommended, however, if one is interested in the absorption pattern of the crystal structure of the solid compound.

The most extensive information about the crystal structure is obtained if one can prepare a very *thin plate*, 2 by 10 mm and about 10 μm thick. Even polarised light can then be used. Unfortunately this technique is restricted to a few cases only. In general no such crystals can be produced.

The *film technique* is applicable in several cases. The compound is dissolved in a volatile solvent. The solution is applied onto an NaCl or KBr window in drops after which the solvent is evaporated. A clear transparent film may be the result. Fatty compounds can be easily handled in this way. Most compounds however produce opaque films. For low-melting materials the preparation of a film via the liquid state can be tried between two windows. Films generally have a polymorphic or amorphic structure.

KBr is transparent up to about 28 μm. KBr powder, mixed with a few mg of the unknown solid, can be made transparent by increasing the pressure to about 10000 kg cm^{-2} and moulding a disk of the mixture under this pressure.

The concentration in a *KBr disk* should be about 0·3%. Although this technique would seem to be ideal, there have been found to be serious drawbacks in practice. One has to be prepared for phenomena such as polymorphism, chemical (exchange) reactions, sensitivity to moisture, adsorption etc.

Furthermore, the technique requires some experience before useful transparent disks are obtained.

Of course numerous variations are possible with these techniques. One has to bear in mind that in fact each problem demands its own technique, and so one has to be familiar with the advantages and disadvantages of all possible ones.

POLYMORPHISM

The occurrence of several crystal forms of a compound is called polymorphism. Each form will show its own absorption pattern, differing slightly or significantly from the others. As this might give identification problems many workers prefer the solution technique. The molecules in a solvent will be 'unaware' of their origin and thus will give rise to one spectrum only. The disadvantages are clear: the problems concerning the solvent and the loss of information about the crystal structure. The KBr technique overcomes both drawbacks at once, but unfortunately may introduce other new problems. It is found that several compounds show alterations in the crystal lattice as result of the grinding procedure and/or the applied pressure. Changes frequently occur in organic materials having a low melting point ($< 100°$C), the compound being rendered completely amorphous. As one does not know beforehand the behaviour of the substance during preparation, one has to be very attentive to polymorphic phenomena. As long as only one form is present each time – no matter what changes in the procedure – there is no need to discard the KBr technique. A run using a mull can lead to a decisive answer. Polymorphism has been observed for benzil, succinimide, several barbiturates, diacetamide, steroids etc.

SOLVENT EFFECTS

The spectra of the same compound in different solvents usually show differences. These are caused by the interaction of the solvent molecules (excess) and the molecules of the sample, and are therefore called solvent effects. Bands may be shifted to other wavenumbers and intensities can also be altered. The frequency shifts may be considerable; a shift of 100 cm^{-1} for a carbonyl group in going from one solvent to another is not uncommon. Many theories have been developed to describe the interaction and to predict the shifts, but so far none really fits the facts. The polarity of the solvent seems to play an important role.

Comparison with reference spectra of solutions is difficult for this reason, without taking into account the solvent's own absorption regions. Structure elucidation is thus hampered by the solvent effect. The conclusive proof of an unknown and a reference spectrum being the same is only possible if all scanning conditions including the solvent are fully identical.

Hydrogen bonding

The attractive force existing between a slightly positively charged hydrogen atom on one hand and a rather negatively charged atom on the other hand is called hydrogen bonding. The hydroxyl group of methanol and the carbonyl group of acetone can be taken as an example:

$$CH_3—O\diagdown_{H\cdots O=C}\diagup^{CH_3}_{CH_3}$$

The hydrogen bond or bridge is represented by the dotted line. The —OH group is the proton donating group, the C=O group is called the proton acceptor. Another proton donator is the —NH$_2$ group. Other proton accepting groups are the halogens, —N(CH$_3$)$_2$ etc.

The attractive force of the bridge is small in comparison with covalent bonds; its absorption bands are to be found beyond our scope in the far infrared. Nevertheless in the normal i.r. region the results of the bond are readily found. It appears that in the formation of the bridge both of the other bonds are involved. In our example the O—H stretching vibration as well as the C=O are shifted to lower wavenumbers.

Not only can a bridge be formed between different groups, but also between identical molecules. For instance, two alkanol molecules may form a dimer:

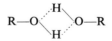

Trimers, tetramers and polymers are possible as well. If one has simultaneously four different types of bridged molecules, then the spectrum will be composed of four different spectra, each belonging to one of the polymers that are present. A very complex spectrum would be the result, but fortunately the hydrogen bond only influences the vibration of the donor and acceptor group and hardly any of the rest of the molecule.

Usually hydrogen bonding, especially O—H bridges, can be easily detected from the specific shape of the bridged group (see p. 76). Instead of the sharp band belonging to the unbridged (free) OH group one gets a rather broad band at much lower frequency. The broadness of the band is due to the fact that the bands belonging to the different polymers lie so close together that they cannot be resolved. Even under high resolving power the bands will not be separated, for they overlap inherently. The differences between the trimer, the tetramer and the other higher polymers are too small to be seen. The free OH and the dimer will, however, show well-separated bands.

Intermolecular bridge. As long as the hydrogen bridge is the result of two molecules attracting each other the problems of interpretation of a polymer spectrum can be avoided by working in a dilute solution. For as one is concerned with equilibria such as

$$n\,\mathrm{ROH} \rightleftharpoons (\mathrm{ROH})_n$$

where $n = 1, 2, 3 \ldots$ it is obvious that in a very dilute solution only the 'free' ROH molecule will be present, the spectrum being then rather simple. The lower the concentration, however, the larger the path length of the cell, and thus the greater the distortion of the spectrum caused by the solvent.

The concentration at which only 'free' molecules are present depends of course on the equilibrium constant; the larger the K the smaller the concentration. This effect may be present to such an extent that the rest of the spectrum is completely obscured by the solvent's own absorption. Yet some bands, e.g. a free OH band, can nearly always be located, since there are several solvents available, e.g. CCl_4, CS_2, which do not absorb in the OH stretching region, even at path lengths up to 10 mm.

Intramolecular bridge. In several molecules the donating and the accepting group are both present, for instance *o*-chlorophenol:

(a)   (b)

Since the bridge is part of the molecule itself, it is called an intramolecular bridge. Contrary to what was said about the intermolecular bridge, diluting the solution will have no effect here. The bridge is permanent, there being no equilibrium similar to that for the intermolecular situation. However, an equilibrium does exist in this case, as part of the molecules will have the same configuration as molecule (b). The bridge has been broken in this case as result of kinetic energy. So in fact there must be an equilibrium constant depending on the temperature. The weaker the bridge and the higher the temperature, the higher the number of 'free' molecules. The equilibrium constant for the intermolecular bridge is also dependent on temperature; the lower it is, the greater the number of molecules in the bridged form.

In the case of an intramolecular bridge dilution may have some effect. This is easily understood if one assumes that at higher concentration an intermolecular bridge is present as well:

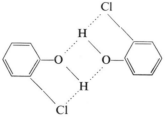

This is possible provided the O----H bridge is weaker than the H----Cl bridge, otherwise no intramolecular bridge would have been detected in dilute solutions.

Whether a hydrogen band has inter- or intramolecular character can be traced by running spectra of dilute solutions. If the band belonging to the bridge is sensitive to dilution the bonding is intermolecular, and at low concentrations a sharp 'free' OH band will appear. If this does not happen, intramolecular hydrogen bonding may be present as well. Alterations in temperature may give further evidence. It must be emphasised that the choice of the solvent is very important in hydrogen bridge measurements. Interaction of the compound under survey and the solvent must be considered to be fully absent. The solvent has to be very pure, and any traces of water, a strong hydrogen bonding agent, must be totally removed.

STRUCTURAL ISOMERISM

Structural isomerism is frequently met in infrared spectroscopy when dealing with liquids, solutions or gases. Consider as an example the molecule 1,3-dichloropropan-2-one. Rotation about the C—C axes is possible provided energy barriers of a few joules/mole can be overcome. Kinetic energy at room temperature is ample for this purpose. Therefore at least three conformations of the molecule are possible:

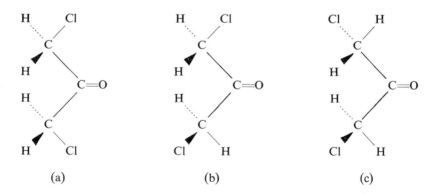

In (a) both chlorine atoms and the carbonyl group are in the C—C—C plane. In (b) just one chlorine atom is in that plane while that of (c) has no chlorine atoms in it. Since those conformations are optically different three types of spectra will arise. Although many normal vibrations will be identical for the three isomers (absorption will take place at the same frequency) some will be clearly different, giving rise to bands at other frequencies. In this case, for instance, three carbonyl bands are found at 1755 (a), 1742 (b) and 1728 cm^{-1} (c) respectively, indicating interaction between the chlorine atoms and the C=O group.

The relative distribution of the isomers will depend to a large extent on the energy barriers that have to be overcome. The intensity of the bands might give some idea of this distribution. A change in temperature will in general alter the ratio, while solvent changes might cause dramatic shifts, both in frequency as well as intensity, interaction with the solvent being the underlying factor.

Sometimes the heights of the energy barriers are such that no equilibrium at room temperature exists; the different isomers can then be isolated. For instance this is the case with many steroids. Infrared spectroscopy can be an effective tool in distinguishing between the different isomers. The spectra of the eight isomers of 5α,β-pregnane-3α,β-17α,β-diol proved to be clearly different from each other. An unknown compound may be readily identified that way. Whilst it is true that structural isomers interfere with the infrared technique, it is thanks to that very interference that infrared spectroscopy can give information about conformational problems.

CHRISTIANSEN EFFECT

The Christiansen effect is restricted to solid samples only. It is caused by a significant difference between the refractive index of the sample and the surrounding material, such as KBr, in the region of an absorption band.

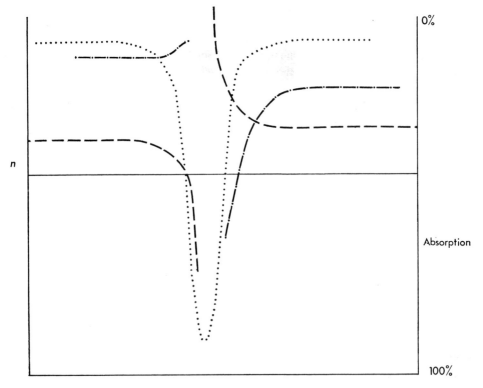

FIG. 35. The change in the scattered light and refractive index when passing an absorption band. - - - - - - - - 'True band', —·—·—·—· scattered light, — — — n_{sample}, ———— n_{KBr}

Suppose $n_{\text{KBr}}^\lambda < n_{\text{sample}}^\lambda$ (which is very common). In the region of an absorption band of the sample n_{KBr} will be fairly constant, but the refractive index of the sample changes dramatically when passing the band (see Fig. 4). As the amount of radiation lost by scattering is proportional to the second power of the difference ($n_{\text{KBr}} - n_{\text{sample}}$), the background line will have the shape as indicated (Fig. 35). Compared with the 'true' absorption band (dotted line) the apparent band will be a distorted one (Fig. 36); a shift in the maximum is very likely. This is known as the Christiansen* effect. It

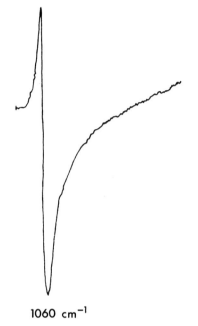

1060 cm^{-1}

FIG. 36. Christiansen effect for the CH band in iodoform (CHI₃)

may be eliminated or reduced by finely powdering the sample until the particles are much smaller than the wavelength of the light being used.

Instead of grinding the sample and the KBr, other mixing techniques may be applied (adding a few drops of a volatile solvent; freeze-drying) to overcome the problems of the particle size.

ORIENTATION EFFECT

When a crystal sample, or a film in which the molecules are orientated, for instance polyethylene, is rotated through a few degrees perpendicular to the radiation beam, significant changes in the intensities of the bands may be observed. Some bands more or less disappear; others show enhanced intensities. Two factors contribute to this phenomenon, called the orientation effect: the instrumental polarisation and the orientation of the molecules with respect to the polarised beam. Parallel bands (vibrations in which the variation of the dipole moment is parallel to the polarised beam) will absorb strongly while the opposite holds for perpendicular bands. Hence it is clear that rotation of the sample in the beam will cause changes in its spectrum.

* W. C. Price and K. S. Tetlow, *J. Chem. Phys.* **16**, 1157 (1948). [Christiansen, *Wied. Ann.* (1884).]

Only a complete distortion of the orientation of the sample molecules can nullify the effect.

In the mull technique this phenomenon is usually absent. Though all the small particles are themselves orientated with respect to the radiation, the overall effect will be zero due to the random distribution. Sometimes, however, the particles are orientated somewhat as result of mechanical forces and/or gravitation. The effect, though small, can then be observed.

QUANTITATIVE ANALYSIS

Introduction

Those expecting infrared spectroscopy to be a powerful method in quantitative analysis will be disappointed. Ultraviolet spectroscopy is far more useful in this respect, but infrared has some important features that make it widely applied. U.v. is about a thousand times more sensitive than i.r. but, unlike i.r., is non-specific. Obviously the analysis of pure compounds is done by u.v. (provided the sample is u.v. active!) whereas mixtures can best be analysed by i.r. Due to the numerous absorption bands of the different compounds some bands specific for each component can nearly always be found. Despite many problems such as the choice of a solvent, the concentration, the solubility, the path length, the background or base line etc., the technique should be used as long as no other or better one is available.

Beer's law

The theoretical relation between the amount of light from a monochromatic beam that passes through an absorbing medium and the amount of absorbant present is given by the Lambert-Beer-Bouguer law, often called simply Beer's law:

$$I = I_0 \exp(-kcl) \tag{1}$$

where I_0 is the intensity (or energy) of the radiation incident on the absorbant, I the transmitted intensity, c the concentration of the absorbant, l its thickness (path length) and k a conversion constant. The formula can be transformed into:

$$\ln I/I_0 = -kcl \quad \text{or} \tag{2a}$$
$$\log I_0/I = \varepsilon cl \tag{2b}$$

where $\varepsilon = k \log e$.

The law is restricted to monochromatic light and for non-interacting absorbants, and hence for very dilute solutions. In practice however it turns out that small deviations from these conditions are possible without serious drawbacks.

Frequently E (extinction), O.D. (optical density) or A (absorbance, IUPAC notation) is used for $\log I_0/I$, thus leading to the linear relation:

$$A = \varepsilon cl \tag{3}$$

A plot of A versus c keeping l constant will be a straight line. Substituting $c = 1$ mmole/ml and $l = 1$ cm one obtains

$$A = \varepsilon$$

and since A is dimensionless, ε will be in $cm^2/mmole$; it is often referred to as the molar absorption coefficient. It may vary largely from peak to peak but in general it lies between 0 and 300 $cm^2/mmole$.

Transmittance and extinction

Most spectrophotometers produce spectra linear in per cent transmittance ($\%T$). Since $T = I/I_0$ eqn. (3) can be changed into

$$A = \log(100/\%T) = \varepsilon cl \qquad (4)$$

$$\text{e.g.} \quad \%T = 50 \quad A = 0{\cdot}30$$
$$\%T = 10 \quad A = 1{\cdot}00$$

T will vary between 100 and 0%, corresponding to a variation in A from 0 to ∞. For a background that does not coincide with the 100% transmittance line, a normal situation, the calculations seem to be more complex, but in fact are as simple as before.

Suppose the background is found at T_1, the peak maximum at T_2. The A for the band is A_2 corrected for the background, or mathematically

$$A_{\text{band}} = A_2 - A_1 \qquad (5)$$

Substituting A_2 and A_1 one gets

$$A_{\text{band}} = \log(100/\%T_2) - \log(100/\%T_1) \qquad (6a)$$

or

$$A_{\text{band}} = \log \frac{\%T_1}{\%T_2} \qquad (6b)$$

Let us conclude this section with an example that is often a source of misunderstanding. A doubling of the concentration or the path length does not halve the band intensity!

Consider a band varying from 100 to 50%. As we saw before $A = 0{\cdot}3$. A doubling of the concentration will lead to $A = 0{\cdot}6$. This means that $\log(100/\%T) = 0{\cdot}6$ or that $T = 25\%$. So the band intensity is only increased by 25%!

Extinction and accuracy

A further look at the relation between T and A shows that there is a most sensitive and accurate range for A as well as very inaccurate ranges. This can be seen in Table 2, where A is calculated assuming that T values are accurate within $\pm 1\%$ absolute.

TABLE 2

$\%T \pm 1$	$A \times 10^3$	Percentage accuracy
99	$4{\cdot}5 \pm 4{\cdot}5$	± 100
90	46 ± 5	11
80	97 ± 6	6
70	155 ± 6	4
60	222 ± 7	$3{\cdot}2$
50	301 ± 9	$3{\cdot}0$
40	398 ± 11	$2{\cdot}8$
30	523 ± 15	$2{\cdot}8$
20	699 ± 22	$3{\cdot}1$
10	1000 ± 46	$4{\cdot}6$
2	1699 ± 238	14

The smallest error is made for A values between 0·4 and 0·5, corresponding to 40 and 30%T respectively. In general, the range between 60 and 30% transmittance is considered to be satisfactory for quantitative analysis. Of course, this type of error is not the only one, but we have mentioned it here as it is a specific one, the result of the inaccuracy of the instrument and Beer's law.

Slit width and true absorption band

Not only the resolution (*cf.* p. 41) but also the shape of a band is sensitive to the slit width. The true absorption band is the curve that one would obtain if it was possible to measure the absorption at each wavenumber $\sigma_1, \sigma_2, \sigma_3 \ldots \sigma_n$, i.e. with

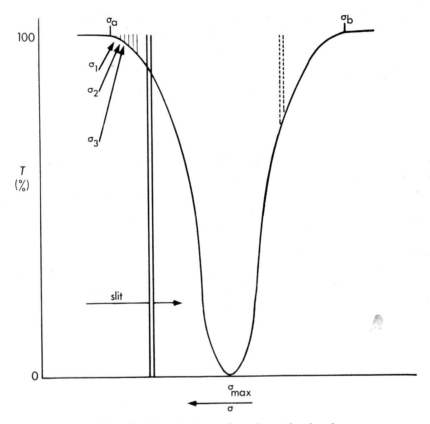

FIG. 37. Normal shape of an absorption band

monochromatic light. In general the band will have a Gaussian or Lorentzian form and the absorption for $\sigma_1, \sigma_2, \sigma_3 \ldots$ etc. is seen by dividing the curve in infinitely small segments of wavenumbers $\sigma_1, \sigma_2, \sigma_3 \ldots$ (see Fig. 37).

Suppose a slit could be set such that monochromatic light came through (unfortunately it is impossible to do so as diffraction phenomena occur). Varying the wavenumber of this beam and measuring the quantity (percentage) of absorbed light for each wavenumber will produce the true band. In fact this does mean, however, that the 'slit is passing the true band' (Fig. 37).

For non-monochromatic light passing a slit, i.e. the real situation, the deformation of the true band can now be predicted. First, say two frequencies σ_p and σ_q pass together; the average wavenumber $\sigma_n = (\sigma_p + \sigma_q)/2$ is indicated on the spectrometer or in the spectrum (see Fig. 38).

For $\sigma_q = \sigma_a$ the absorption starts, but σ_n is still greater than σ_a and for $\sigma_p = \sigma_b$ the absorption ends, but here $\sigma_n < \sigma_b$ and hence the band is broadened. For $\sigma_n = \sigma_{max}$ the absorption will have an average value that will be always smaller than the true one.

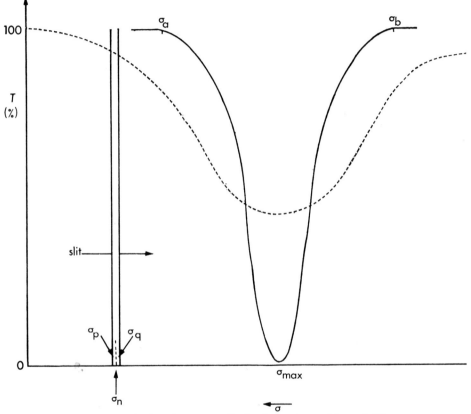

FIG. 38. The band contours for both narrow and broad slits

In practice, such small deviations from monochromaticity hardly influence the band contour. For broader slits (many wavelengths at once), however, the situation is worse. The deviation from the true band will be considerable; the apparent peak being broad and less intense (see Fig. 38). Hence it is clear that the extinction depends to a large extent on the slit width; very unfortunate for quantitative work. From practice one knows that as long as the spectral slit width is smaller than one-fifth of the half bandwidth of the true band this band will still be produced.

The spectral slit width $\Delta\sigma_{\frac{1}{2}}$ is defined as the difference between the two extreme wavenumbers σ_m and σ_n at half-height in a plot of the distribution of energy passing the slit against the wavenumber with the instrument set at a fixed wavenumber (see Fig. 39).

The half bandwidth of the true band is defined in a similar way. If this band is very sharp it will be impossible to get satisfactory reproduction as the slit cannot be set so narrow. A distorted band will be the result and quantitative work without a standard curve may then be a difficult job.

It is found, however, that the band area, i.e. the integrated absorbance

$$\int_{\sigma_1}^{\sigma_2} A_\sigma \, d\sigma$$

is more or less independent on the slit width and is thus a more suitable value for quantitative work. Integration can only be done for $A - \sigma$ curves and not for $\%T - \sigma$

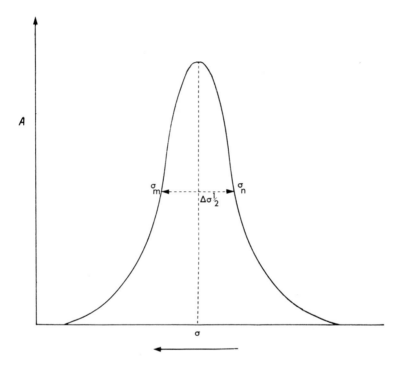

FIG. 39. Energy distribution as a function of the wavenumber for the radiation passing the slit at an indicated wavenumber σ

curves. These have first to be transformed into $A - \sigma$ which is a time-consuming job. For isolated bands the choice of σ_1 and σ_2 is usually simple.

For quantitative work on different spectrometers the integrated absorption is to be preferred. For normal work using just one machine the $A_{\max}$ value will do; this saves a lot of calculation and measurement.

Band choice

A few remarks are necessary concerning the choice of a band upon which the calculations will be based. The band should be:

(a) strong: the stronger it is, the lower the detection limit(s).
(b) broad: it must not be sharp, as a very small deviation from the wavenumber maximum will result in a serious error in A.

(c) isolated from other bands, otherwise interference with those bands may occur.

(d) outside regions where compensated absorption occurs. In addition to solvent absorption there is also that due to atmospheric carbon dioxide and water vapour.

(e) free from any effects such as there are in hydrogen bonding, dissociation, etc. except if measurements on these phenomena are to be obtained.

If one or more of these requirements cannot be fulfilled, one will have to be satisfied with a less accurate determination of the concentration, though obviously this is better than none at all.

Base line

For completely isolated bands the background line, i.e. the line obtained under fully identical conditions but without a sample, can easily be drawn. Often the peak being measured is anything but isolated. A band as represented in Fig. 40 is not rare. In this

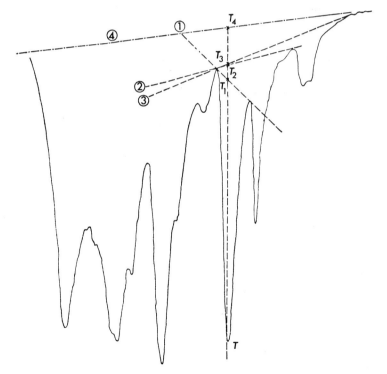

FIG. 40. Probable base lines for an absorption band

case a base line is used instead of the unknown background line. It is obvious that several lines can be chosen in this way. The best fit will be found from practice when a A-c plot is prepared. There is no particular reason why one should be better than another. Hence there is no need to look for the 'real' base line. As long as the calibration line, A versus c, seems to be normal, the choice of the base line is arbitrary.

Internal standard

If for reasons of practice the path length is unknown (mull, KBr technique, uncalibrated cells), an internal standard may be added to the sample. This standard should only have a few bands and at least one intense band free from any other (e.g. KCN, where the C≡N band is used). The standard is added to the mull, the KBr or the solvent in an exactly known quantity beforehand. The extinction of the unknown sample is now determined with respect to the extinction of the standard. The path length cancels out this way. For instance when the path length is doubled both A's will be doubled, but the ratio will still be the same, indicating that the concentration is unchanged.

Here too a calibration line, $A_{sample}/A_{standard}$ versus c_{sample}, has to be made in advance. The method is rather complicated and moreover it is doubtful whether the mull and KBr techniques are appropriate at all for quantitative work.

Cell calibration

For the calibration of the path length of a cell the interference fringe method is used. A parallel monochromatic beam perpendicular to the cell windows is passed through the cell. Part of the beam, however, is reflected backwards at the four surfaces of the window material. The light passing the rear surface of the first window will interfere with the light reflected by the front of the second window. Dependent on the wavelength and the path length, there will be destructive interference if $2d = (n + \frac{1}{2})\lambda$ where d is the cell thickness in μm, λ the wavelength in μm and n is 1, 2 . . . any integer.

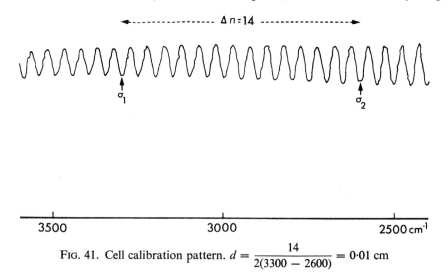

$$d = \frac{14}{2(3300 - 2600)} = 0.01 \text{ cm}$$

FIG. 41. Cell calibration pattern.

Varying λ continuously will give rise to the appearance of an interference pattern (see Fig. 41), from which d can be calculated using the following formula:

$$d = \frac{\Delta n}{2(\sigma_1 - \sigma_2)}$$

where σ_1 and σ_2 are the wavenumbers between which the number of fringes, Δn, is counted. This method will only give good results provided the windows are flat and parallel to each other. Otherwise distorted patterns or no pattern at all result.

Attenuated total reflection

When a light beam in a medium of high refractive index is reflected at the interface with a medium of a lower index, an evanescent wave is set up in the latter medium propagating parallel to the interface for a very short distance. Eventually the wave returns to the first medium, making the reflection total. This phenomenon can be fully understood from Maxwell's law for electromagnetic radiation.

Now suppose that the evanescent wave when propagating in the second medium is slightly absorbed. The reflection will then no longer be total but attenuated. This is

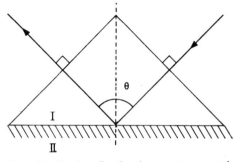

FIG. 42. Single reflection in an a.t.r. crystal

called attenuated total reflection (a.t.r.). This attenuated beam can give us information on the second medium.

The use of a.t.r. in infrared spectroscopy started in 1960 with Fahrenfort's* and Harrick's† papers on theory and practice. The basic idea of the measurement is given in Fig. 42.

A ray or beam perpendicular to medium I is reflected at the interface with medium II as result of the difference in refractive index. The angle of incidence θ is greater

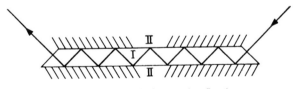

FIG. 43. Multiple internal reflection

than the critical angle. The ray comes out again perpendicular to the polished surface.

Since the penetration depth of the reflected beam (or more correctly, the evanescent beam) is approximately its wavelength, the effective path length is small (a few μm) and thus the absorption will also be small. This can be improved simply by increasing the number of reflections. The principle of such methods is given in Fig. 43. It is commonly referred to as multiple internal reflection (m.i.r.). The path for one ray through medium I is indicated. The number of reflections can be altered by changing the angle of incidence. Obviously more sample (medium II) is needed to cover medium I on both sides.

* J. Fahrenfort, *Spectrochim. Acta* **17**, 698 (1961); **18**, 1103 (1962).
† N. J. Harrick, *Phys. Rev. Letters*, **4**, 224 (1960).

Several a.t.r. units fitting most commercial spectrometers are now available. The principle of m.i.r. has been widely accepted. We will mention briefly the disadvantages, limitations and, more important, the advantages of this technique for i.r. work.

The a.t.r. material (medium I) is a critical component. The desirable characteristics are well known, but hard to fulfil. The material should be chemically inert (no reaction with the sample), pure (no absorption), quite tough and easy to polish and, last but not least, the ratio $n_{material}/n_{sample}$ should be greater than unity, otherwise there will be no reflection at all, although it should not be too high, as the attenuation is inversely proportional to it. Well-known a.t.r. materials are KRS-5, germanium, AgCl or AgBr and silicon. Others may be chosen for special purposes.

Restrictions as to the sample are less stringent. In general all materials that can make good contact with the a.t.r. plate will do. If the peaks are too strong, the sample area can be diminished for better results. Powders and rough samples will cause difficulties, as can be appreciated.

The only drawback of the a.t.r. technique results from the a.t.r. accessory itself. It causes a lengthening of the light path of the sample beam (the double beam character becomes lost) and a loss of light.

On the other hand, a.t.r. has advantages over the transmission techniques, especially for opaque materials. The method can be applied to different samples such as polymer films, coatings, fabrics, paints, pastes, leather and many others. Sampling itself is perhaps even simpler than that for the transmission method.

The intensity of a band in an a.t.r. spectrum depends on (a) the penetration depth into the sample and (b) the ratio of the refractive indices. Both vary with the wavelength but the latter can show dramatic changes when 'nearing or passing' an absorption band (cf. p. 54). For that reason part of an a.t.r. spectrum may look like a transmission one (the resemblance may be striking) but the overall appearance will be different.

Quantitative work is hardly possible unless only moderate or low accuracy can be accepted. The biggest problem is the reproducibility in mounting the a.t.r. plate and the sample, as the effective area of the sample must remain constant.

Micro attachments are available now, and these sets will make the use of the a.t.r. technique still more popular.

BIBLIOGRAPHY

R. G. J. Miller and B. C. Stace, *Laboratory Methods in Infrared Spectroscopy* 2nd edn., Heyden & Son, London, 1972.
L. May, *Spectroscopic Tricks*, Plenum Press, New York, 1967.

5

Interpretation of spectra

The interpretation of spectra, i.e. correlating spectra with molecular structure, is in many ways an art. Fortunately, some general rules can be outlined to facilitate learning this art, although some feeling and a lot of experience are required to be able to do it really well.

First let us recall some results from theory. In a spectrum two rather different types of absorption bands are present: those attributable to distinct parts of a molecule ('group frequencies') and those caused by vibrations of the molecule as a whole ('skeletal modes').

The group frequencies are localised at certain regions in the infrared spectrum and, as can be appreciated, are very useful indeed for the identification of the functional groups of a molecule. The skeletal frequencies are characteristic of a particular molecule and as such are not localised at all, though in general they do occur below 1500 cm^{-1}. (This region, therefore is often referred to as the 'fingerprint' region.)

From the preceding it may be clear that the interpretation of a spectrum will have to start with the identification of the more or less localised group frequencies.

A functional group can give rise to none, one, or more than one absorption band, depending on the nature of the group. For instance a symmetrical $\diagup C{=}C\diagdown$ system will not absorb any radiation, the transition being forbidden in the infrared. A C$\equiv$N group leads to one and only one band (the C$\equiv$N stretching vibration) in the 2200 cm^{-1} region, whereas a $\diagup CH_2$ group will give rise to at least 6 bands as we saw before (see Chapter 2).

Let us now apply the above principles, taking the C$=$C group as an example. The band is to be found in the 1650 cm^{-1} region.

(a) No C$=$C band in the spectrum. Three possibilities arise:

1. The molecule lacks this functional group.
2. The group is present but the transition is infrared inactive.
3. The transition is allowed, but the intensity of the band is too small to be seen under the applied conditions.

(b) One C$=$C band in the spectrum. Again three possibilities:

1. Only one C$=$C group is present in the molecule.
2. Two or even more fully identical groups form part of the molecule; the bands coincide.
3. As under 1 or 2 but other inactive C$=$C groups are present.

(c) Two or more C=C bands are found in the spectrum. Two possibilities arise:

1. At least two *optically* different types of C=C bands are present in the molecule.
2. As under 1 but also considering the foregoing possibilities (e.g. inactive groups may form part of the molecule).

The same holds for other groups except that there are many groups without forbidden transitions such as C≡N, C=O, etc.

For functional groups of the complex type (CH_2, CH_3, NO_2, NH_2, etc.) the situation is similar. Instead of one band several bands will appear at different regions in the spectrum, each band indicating the presence of the group. The bands are not independent and would thus seem to be superfluous or at least useless for interpretation. These bands are far from useless, however, as we will see later on.

To return to the functional group regions, how do we know the region(s) where a certain functional group will absorb radiation? This is not as difficult as it looks. From the early days of infrared spectrometry data has been compiled on this subject. These data are tabulated in numerous tables (see Appendix D), but they can also be presented on correlation charts (see Appendix C). On such charts the group frequency regions are indicated by lines. Functional groups showing more bands are thus represented by more lines. The length of a line denotes the broadness of the region.

When looking over the correlation charts one can see that many regions overlap showing that different functional groups can show absorption bands that coincide. Though it is obvious that the identification of a certain band is hampered that way, one can make use of the earlier mentioned extra bands. For instance, a CH_3 group absorbs at 2900 cm^{-1} and so does a CH_2 group. Since a CH_3 group will have also a band at 1350 cm^{-1} and a CH_2 group will not, evidence is found for one of the two possibilities. Actually the situation is more complicated than this usually, and there are usually more than two possibilities.

So far, only the frequency (wavenumber) of a band has been considered. The intensities can give new or additional information. A rough idea on the intensity can be obtained from this rule-of-thumb: the higher the dipole moment involved in the vibration of the functional group, the higher the intensity of the band. Carbonyl (C=O) bands are strong, aromatic C—H bands weak, whereas —C≡C— bands have variable intensity, generally speaking. It will be clear that the presence of more identical groups in the same molecule alters the situation. For instance a C=O group in one of the high alkanones will show only medium intensity. Furthermore the preparation technique has not been taken into account; concentration differences as well as scanning conditions can change the intensity of a peak. Obviously the use of the band intensity rather than its frequency is a difficult task, not least as a result of the arbitrarily chosen terms weak, medium, strong, etc. Experience is an important factor in this field.

A third parameter useful in the interpretation of functional groups is the shape of a band. Although the shape depends on the intensity and moreover is highly influenced by the representation of the spectrum – linear in wavelength or wavenumber, scanning conditions – some functional groups can be recognised at once by their shapes while the choice is often harder to make in other cases. A few examples of some characteristic absorption patterns are given in Appendix B.

As to the tables and the charts found in the literature and here in this book, some preliminary remarks should be made. The tables and thus the charts are obtained by simply compiling data from the literature. In general these data are not critically reviewed before quoting. Scanning conditions are not taken into consideration at all. This is an extremely unsatisfactory situation for the following reasons:

(a) most data in the literature are given without mentioning the accuracy.

(b) data are obtained from spectrophotometers with different resolution.

(c) the spectrophotometers are often insufficiently calibrated.

(d) data that do not match are compared with each other, using different phases, different solvents, etc.

These and other causes significantly lower the value of the tables. This situation will change, but for the time being use must be made of the available data.

The tables in Appendix D are based on data from the literature as well as from the author's own sources. They are presented in a rather condensed form; more details can be found in standard works (Appendices A and E).

We now return to the problem of the unknown spectrum. There are no special rules for the beginner and so he will have to think along the same lines as the expert. Some guidance here may be very helpful. The most important rule is: think logically. Consider all possibilities you can think of, even the rarest ones, and eliminate them one by one if evidence is found for this in the spectrum.

Start with the identification of the functional groups. Do not forget the preliminary data, such as the phase, the colour, the smell or its origin, etc. Proceed by drawing up a provisional molecule and consider whether this molecule would give a similar spectrum or not. If not, start again. If so, try to find a spectrum of this compound in reference collections or anywhere in the literature and compare this with that of the unknown compound. If there is a difference try to find out the reason for that and change the provisional formula. Compare again with a reference spectrum. In case it is not at hand try to find the substance and see by personal measurement what the spectrum looks like. Proceed until two identical spectra – the unknown and the reference – lie on the desk. Only then may the interpretation be called successful.

BIBLIOGRAPHY

A. J. Baker and T. Cairns, *Spectroscopic Techniques in Organic Chemistry*, Heyden & Son, London, 1966.
A. J. Baker, T. Cairns, G. Eglinton and F. J. Preston, *More Spectroscopic Problems in Organic Chemistry*, Heyden & Son, London, 1967.
L. J. Bellamy, *The Infrared Spectra of Complex Molecules*, Methuen, London, 1960.
L. J. Bellamy, *Advances in Infrared Group Frequencies*, Methuen, London, 1968.
K. Nakanishi, *Infrared Absorption Spectroscopy*, Holden-Day, San Francisco, 1962.

Appendix A

REFERENCE SPECTRA: MAJOR COLLECTIONS

Sadtler Standard Spectra

Sadtler Research Laboratories Inc., have for many years been the leading publishers of spectral data. Their collections cover Infrared, Ultraviolet, Nuclear Magnetic Resonance and Differential Thermal Analysis. All Sadtler Standard Spectra are continuing data projects and all spectra are scanned at SRL. Spectra are available printed on paper or in Microfilm as Film or Fiche.

Two Infrared Collections are available: (1) Infrared PRISM Standard Spectra and (2) Infrared GRATING Spectra. The IR Prism collection now totals 41,000 Spectra and by annual subscription a further 2,000 spectra are added each year. The IR Grating collection was started more recently and apart from an issue of 2,000 spectra of new compounds each year an additional 1,000 spectra are issued of compounds which were previously scanned for the prism collection. Subscribers can choose whether they want the full subscription of 3,000 spectra annually or the non-duplicating 2,000 spectra. The IR Grating Collection now comprises 22,000 spectra.

The Sadtler Collections are well supported with excellent indices. These give reference to all Sadtler Standard Spectra which may be available in the IR Prism, IR Grating, Ultraviolet, Nuclear Magnetic Resonance or Differential Thermal Analysis collections. A recent addition is that the 8,000 spectra published by the Coblentz Society, printed and distributed by SRL, are also included in these 'Total' indices. Retrieval is possible with four different Sadtler Total Indices by: (a) Molecular Formula, (b) Chemical Classes (c) Alphabetical Name, (d) Numeric Serial Number.

A most valuable index permitting the identification of unknown spectra is the Sadtler 'Spec-Finder'. With the aid of this Spec-Finder searches of spectra can be made by the positions of absorption bands in the unknown.

Smaller Special Collections of Sadtler Infrared Spectra comprise ATR – Attenuated Total Reflectance Spectra; Biochemicals; Steroids; Inorganics and Organometallics.

Whereas the Sadtler *Standard* Spectra comprise only pure organic compounds, the Sadtler *Commercial* Spectra collections consists of groups of spectra of commercially available products, most of which are available as Prism or Grating Spectra.

These separately available groups of commercial spectra are:

Agricultural Chemicals	Drug and Drug Extracts	Fibres
Coating Chemicals	Dyes, Pigments & Stains	Food Additives

Intermediates	Petroleum Chemicals	Rubber Chemicals
Lubricants	Pharmaceuticals	Solvents
Monomers & Polymers	Plasticisers	Surface Active Agents
Natural Resins	Polyols	Textile Chemicals
Perfumes & Flavours	Pyrolysates	Water Treatment Chemicals

Each IR Commercial Spectra group has its own index. Two additional *Composite* indices listing all commercial Spectra from all groups are also available: (a) The Commercial Spectra Alphabetical and Molecular Formula Index and (b) The Commercial Spectra Spec-Finder.

DMS – Documentation of Molecular Spectroscopy

The second major collection is the DMS System. The spectrum and relevant data are printed on Needle-Sort Cards. Because the System has grown to over 19,000 Spectra the collection has become too comprehensive for this method of retrieval. Whereas the Sadtler spectra are all run at SRL under standard condtions the DMS spectra are collected from different sources, fulfilling, however, certain requirements set by the DMS Editorial Board. The chart presentation is smaller than Sadtler but one can profit from the listing of wavenumbers (or wavelength) which frequently appear alongside the spectrum. Printed lists of indices with supplements are available but DMS Index Cards comprising peep-hole cards aim to simplify sorting for reference spectra. Each Card has a grid for 5,000 punchable positions. The serial number of each spectral card corresponds to one of the positions on the peep-hole card, while each peep-hole card corresponds to a property of the code of which there are 211.

An additional service is the DMS Literature Service which lists all relevant literature on infrared, Raman and microwave spectroscopy that has appeared since 1963. The DMS System is published by Butterworths and Verlag Chemie.

API/TRC – The American Petroleum Institute Research Project 44 and the Thermodynamics Research Center Data Project

Both projects issue initial sets of looseleaf sheets in six categories, revised and updated semi-annually. Both are located at the Thermodynamics Research Center.

The American Petroleum Institute Research Projection 44 (API RP 44) compiles, calculates, critically evaluates and publishes tables of selected physical and thermodynamic property values and selected spectral data in 5 categories for the classes of hydrocarbons and certain classes of organic nitrogen hydrocarbon derivatives of interest to petroleum and petrochemical industries. The collective title of the publications is 'Selected Values of Properties of Hydrocarbons and Related Compounds'. Infrared spectra fall under Category B; there are 3079 spectra currently available.

The Thermodynamics Research Center Data Project (TRC), (formerly The Manufacturing Chemists Association Research Project) publishes similar data broken down into the same categories as API RP 44 in 'Selected Values of Properties of Chemical Compounds' for the classes of organic compounds other than hydrocarbons and for certain other classes of inorganic compounds of interest to the chemical industry. There are 665 infrared spectra currently available in this collection.

IRDC – The Infrared Data Committee of Japan

These cards are also of a needle-sort type. The set now comprises 10,000 cards and a further 1,000 spectra are issued per annum.

Coblentz Society Spectra

This Society collects spectra from various sources and examines and classifies this data. More recently, from spectrum 5,001 onwards the spectra are critically evaluated to conform to minimum standards of Class III, set by the Coblentz Society. Spectra 1–5,000 Selected Spectra; 5,001–8,000 Critically Evaluated Spectra. These 8,000 spectra are printed and distributed for the Coblentz Society by Sadtler Research Laboratories and are now also indexed in the SRL Total Indices. The Spectra are available on paper or in Microform as Film or Fiche.

Infrared Spectra of Selected Chemical Compounds—Microform Edition

A smaller complete collection of about 2,000 spectra recorded linear in wavelength (wavenumber tables supplied) of well-chosen simpler type compounds, invaluable for laboratories that have problems in establishing a large collection, either due to lack of funds or because their work is only in specialised fields. An ideal teaching aid, the spectra were formerly available in a paper edition arranged simply in Serial Number order. The new low-cost edition is available on Microfilm or Microfiche and it is important to note that the spectra have been re-arranged into a Chemical Classes Order which permits easy comparison of the spectra of related compounds. An index provided also permits the retrieval of spectra by Name, Molecular Formula, Chemical Class or Serial No.

Published by Heyden & Son.

REFERENCE SPECTRA: MINOR COLLECTIONS

Bellanato and Hidalgo: *Infrared Analysis of Essential Oils* contains 214 infrared spectra of essential oils which can be compared using a split binding with any of 60 of their constituent compound spectra. Heyden & Son, London, 1971.

Dobriner *et al. An Atlas of Steroid Spectra* contains 760 spectra, Vol. 1, 1953, Vol. 11, 1958. Wiley-Interscience, New York.

Haslam and Willis: *Identification and Analysis of Plastics* contains 300 spectra of plastics and resins. Iliffe, London, 1965.

Holubek: *Spectra Data and Physical Constants of Alkaloids.* A continuing data project. Each spectral sheet shows ultraviolet and infrared spectrum. Issues 1–8 cover Sheets 1–1000 in five binders. Heyden & Son, London.

Hummel and Scholl: *Infrared Analysis of Polymers, Resins* contains 1,758 spectra of polymers, resins and additives. Wiley-Interscience, New York, 1969.

Neudert and Röpke: *Atlas of Steroid Spectra.* Contains 900 spectra of steroids. Springer-Verlag, Berlin, 1965.

Welti: *Infrared Vapour Spectra* (Group frequency correlations, sample handling and the examination of gas chromatographic fractions) contains 300 spectra of volatile organic compounds. Heyden & Son, London, 1970.

JOINT INDICES FOR ALL PUBLISHED INFRARED SPECTRA

ASTM – American Society for Testing and Materials

Three universal indices have been issued by ASTM covering 92,000 Spectra available in any of the collections or in the original literature.

AMD–31 *Molecular Formula List of Compounds, Names and References to Published Infrared Spectra.* This is undoubtedly the best buy for anyone who wishes to locate spectra in collections or in the literature. The book comprises 616 pages and is relatively inexpensive.

AMD–34 *Alphabetical List of Compound Names, Formulae and References to Published Infrared Spectra.* The first section is an alphabetical list of organic compounds with molecular formulae.

The second section is an alphabetical list of organic compounds, where the molecular formulae are unknown. It also includes inorganic compounds in alphabetical order.

The third section again gives cross reference to the original literature in which abstracted spectra appeared.

AMD–32 *Serial Number List of Compound Names and References to Published Infrared Spectra.* It is a companion to the ASTM Spectral Data File, since it identifies a compound from the spectrum serial number obtained by searching this data file.

IRSCOT System

A different approach to infrared spectral interpretation is used by the Miller and Willis IRSCOT System (Heyden & Son). This consists of data cards on each characteristic group frequency together with a correlation table index, permitting rapid access to this data. Each data card contains concise and reliable information on the infrared absorption bands of a structural group and gives appropriate literature references and examples. Band positions are quoted in both frequency and wavelength. The ten sections of the system so far published cover:

1. Hydrocarbons
2. Halogen Compounds
3. Oxygen Compounds (excl. acids)
4. Carboxylic Acids and Derivatives
5. Nitrogen (excl. N-O Compounds)
6. N-O Compounds
7. Heterocyclics
8. Sulphurs
9. Silicon Compounds
10. Boron Compounds
11. Phosphorus Compounds (in preparation)

Provision has been made for additional or replacement cards to be issued. The cards are supplied in handy binders.

Publisher's addresses*

American Society for Testing and Materials,
1916 Race Street,
Philadelphia,
Pennsylvania 19103, U.S.A.

Butterworth & Co.,
88 Kingsway,
London W.C.2, England.

Heyden & Son Ltd.,
Spectrum House,
Alderton Crescent,
London N.W.4, England.

Data Distribution Office

Thermodynamics Research Center,
Texas A & M Research Foundation,
F.E. Box 130,
College Station,
Texas 77843, U.S.A.

Sadtler Research Laboratories Inc.,
3316 Spring Garden Street,
Philadelphia,
Pa. 19104, U.S.A.

* Sadtler Standard Spectra, API/TRC Spectra, IRDC Spectra, Coblentz Society Spectra, and the ASTM Infrared Indices are distributed in Europe by Heyden & Son, from whom further information can be obtained. In all other cases, contact publisher.

Appendix B

TYPICAL BAND CONTOURS

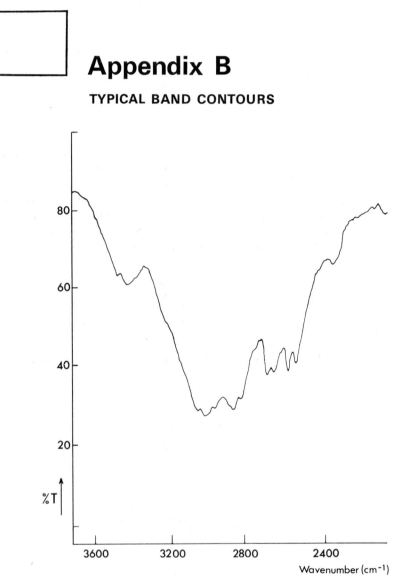

Carboxylic acid (Solid in KBr)

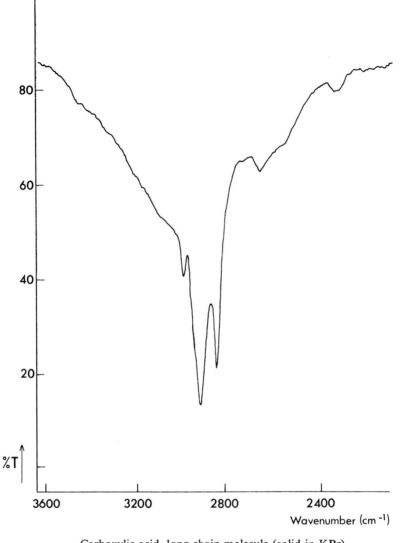

Carboxylic acid, long-chain molecule (solid in KBr)

The broad band is caused by the strongly bridged OH stretching vibration of the carboxylic acid group, on which the CH stretching peaks of the long chain fatty acid are superimposed.

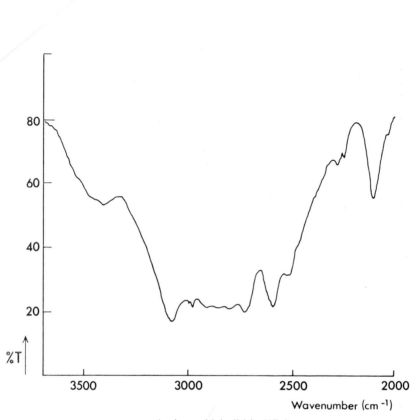

Amino acid (solid in KBr)

This very broad band is caused by the strongly bridged OH stretching vibration of the carboxylic acid group and the nitrogen–hydrogen stretching in the NH_2 and N^+H_3-groups. Besides there may be the peaks of the CH stretching band in the normal regions.

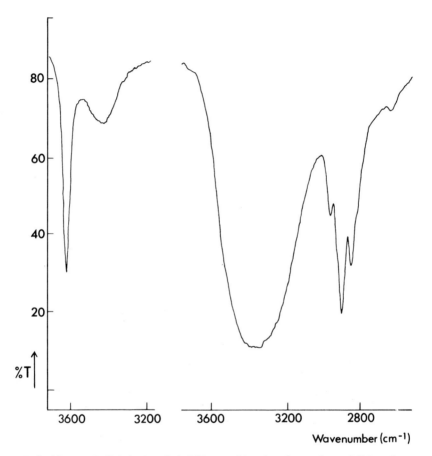

Left: Free and slightly bonded OH stretching bands at about 3600 and 3400 cm^{-1} respectively (solution in CCl$_4$).

Right: The broad OH stretching band of the bridged OH group in an alcohol (pure liquid).

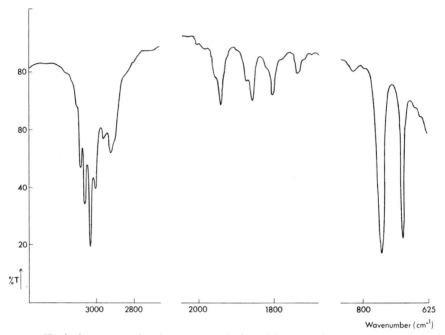

Typical patterns related to mono-substituted benzene rings.

Left: Above 3000 cm^{-1} the CH stretching vibrations of the aromatic
hydrogen atoms.
Middle: Overtone pattern.
Right: Out-of-plane bending vibrations of the 5 adjacent hydrogen atoms
of the aromatic nucleus.

The patterns as such can be different or may even be absent in many cases
especially if hetero atoms such as oxygen are present.

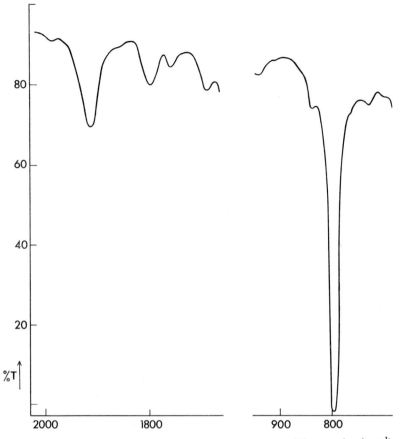

Two regions in the spectrum from which *para* substitution may be recognised (pure liquid).

Left: The overtone pattern, with a dominant peak at about 1900 cm^{-1}.

Right: A strong band caused by the out-of-plane bending vibration of the two adjacent hydrogen atoms attached to each side of the aromatic nucleus.

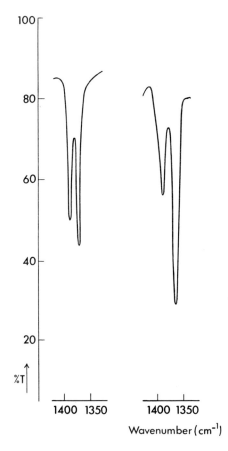

Wavenumber (cm⁻¹)

Left: isopropyl or *gem*-dimethyl group (pure liquid)

Right: tert-butyl group (pure liquid)

The bands caused by the symmetric CH bending vibration of the CH_3 groups will appear if the methyl groups are attached to a saturated carbon atom. The distance between the split peaks is 16 ± 4 cm^{-1} for the isopropyl and 27 ± 4 cm^{-1} for the *tert*-butyl group.

The pattern is frequently distorted by other methyl groups in the molecule.

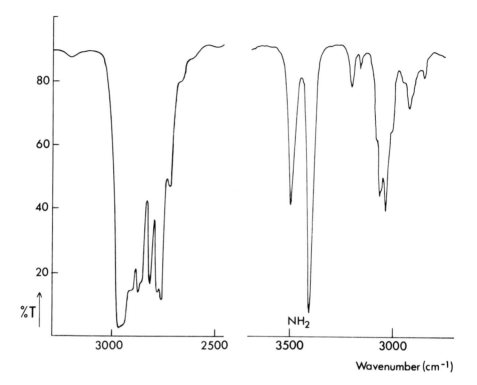

Left: CH stretching region: methyl(ene) groups attached to nitrogen (pure liquid)

Right: NH stretching region of primary amines (solution in CCl₄)

In compounds such as triethylamine, piperidine, dimethylpentylamine, etc. the CH stretching region appears to be extended with peaks at wavenumbers below 2800 cm⁻¹.

The NH₂ group has two bands, i.e. asymmetric and symmetric stretching bands in the region 3600–3300 cm⁻¹ of different intensity.

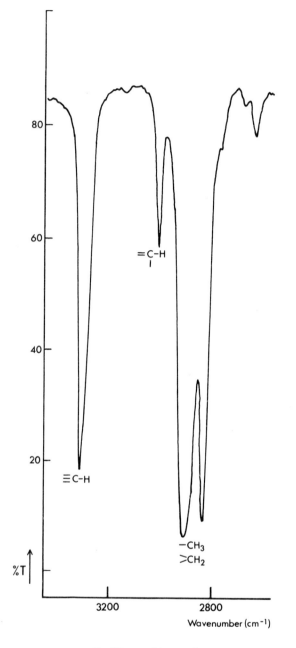

C—H stretching region

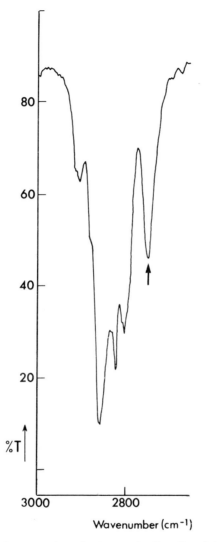

An aldehyde group gives rise to a peak of medium intensity apart from the normal CH stretching region at the low wavenumber side. It appears at about 2700 cm^{-1}. Pure liquid spectrum.

Appendix C

CORRELATION CHARTS

HYDROCARBONS

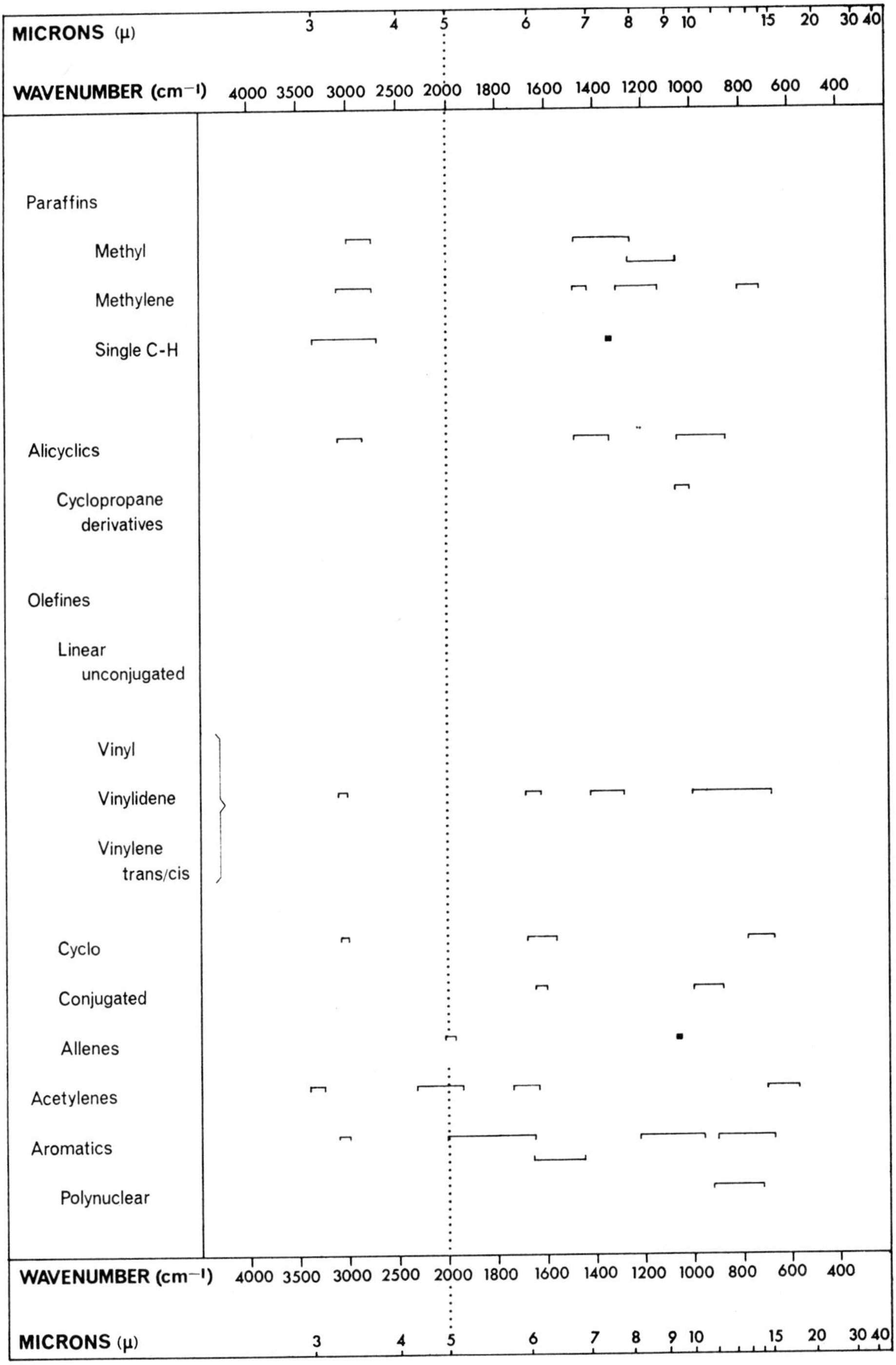

MICRONS (μ)															
WAVENUMBER (cm⁻¹) 4000 3500 3000 2500 2000 1800 1600 1400 1200 1000 800 600 400															

Chart of infrared absorption band positions (MICRONS 3 4 5 6 7 8 9 10 15 20 30 40; WAVENUMBER cm⁻¹ 4000 3500 3000 2500 2000 1800 1600 1400 1200 1000 800 600 400) for the following groups:

Paraffins
  Methyl
  Methylene
  Single C-H

Alicyclics
  Cyclopropane derivatives

Olefines
  Linear unconjugated
  Vinyl
  Vinylidene
  Vinylene trans/cis

  Cyclo
  Conjugated
  Allenes
Acetylenes
Aromatics
  Polynuclear

WAVENUMBER (cm⁻¹) 4000 3500 3000 2500 2000 1800 1600 1400 1200 1000 800 600 400															
MICRONS (μ) 3 4 5 6 7 8 9 10 15 20 30 40															

HALOGEN COMPOUNDS

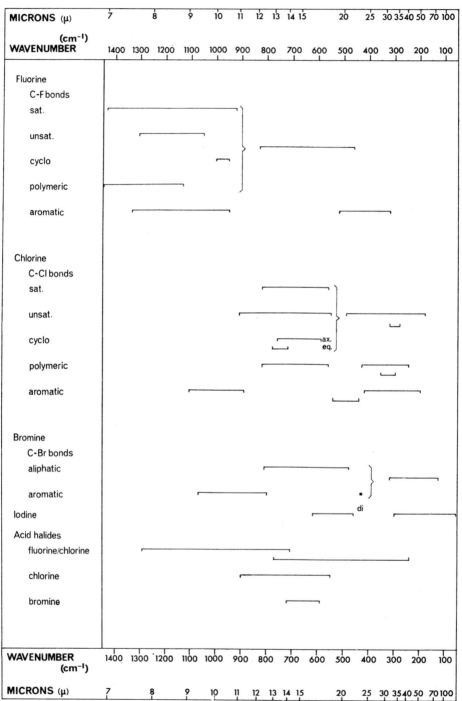

OXYGEN COMPOUNDS
(excluding carboxylic acids)

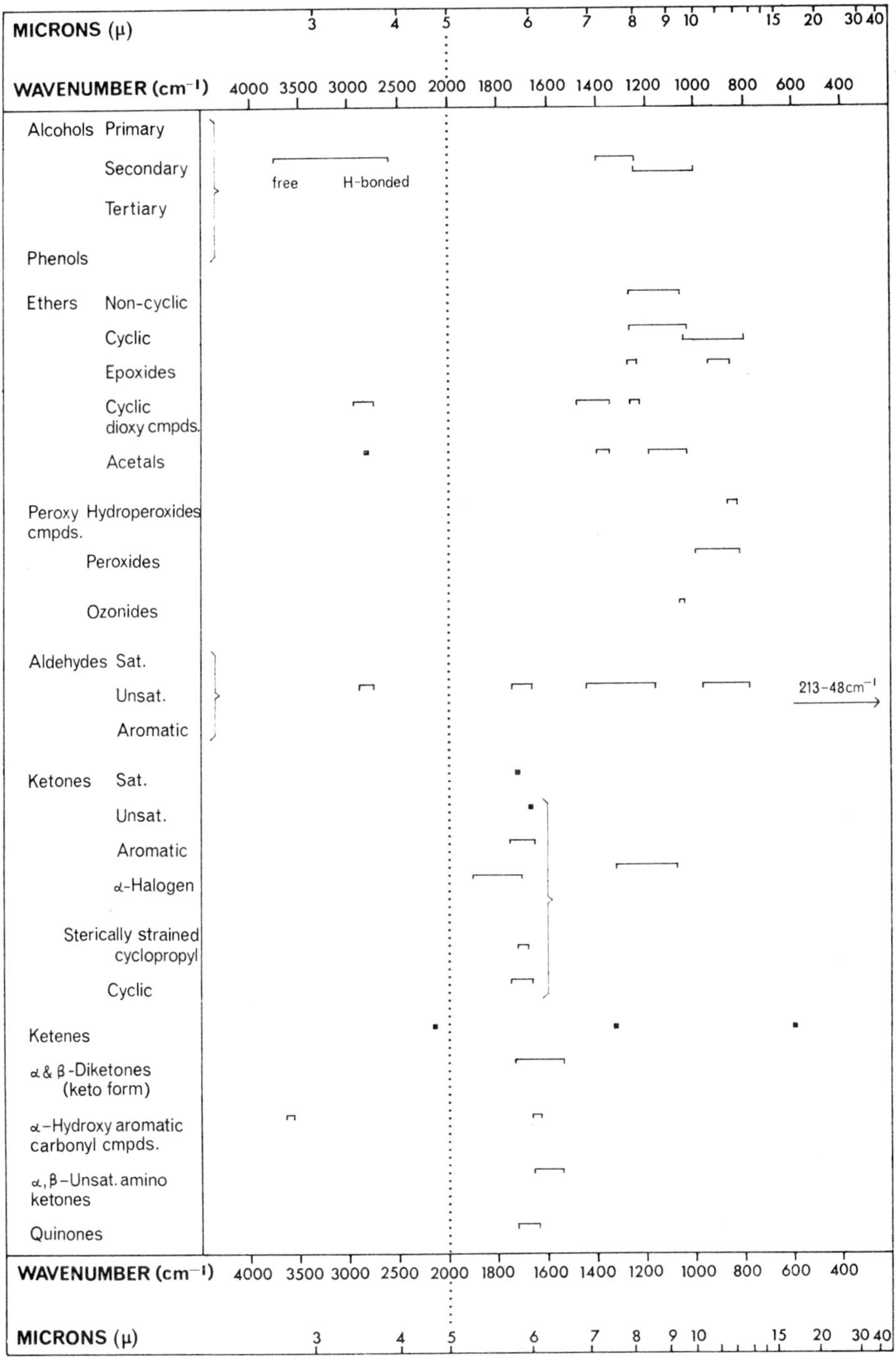

| MICRONS (μ) | | 3 | 4 | 5 | 6 | 7 | 8 | 9 10 | 15 | 20 | 30 40 |

| WAVENUMBER (cm⁻¹) | 4000 3500 3000 2500 2000 1800 1600 1400 1200 1000 800 600 400 |

Alcohols Primary

Secondary   free   H-bonded

Tertiary

Phenols

Ethers   Non-cyclic

Cyclic

Epoxides

Cyclic dioxy cmpds.

Acetals

Peroxy Hydroperoxides cmpds.

Peroxides

Ozonides

Aldehydes Sat.

Unsat.   213–48cm⁻¹ →

Aromatic

Ketones   Sat.

Unsat.

Aromatic

α-Halogen

Sterically strained cyclopropyl

Cyclic

Ketenes

α & β -Diketones (keto form)

α–Hydroxy aromatic carbonyl cmpds.

α,β–Unsat. amino ketones

Quinones

| WAVENUMBER (cm⁻¹) | 4000 3500 3000 2500 2000 1800 1600 1400 1200 1000 800 600 400 |

| MICRONS (μ) | | 3 | 4 | 5 | 6 | 7 | 8 9 10 | 15 | 20 | 30 40 |

CARBOXYLIC ACIDS AND DERIVATIVES

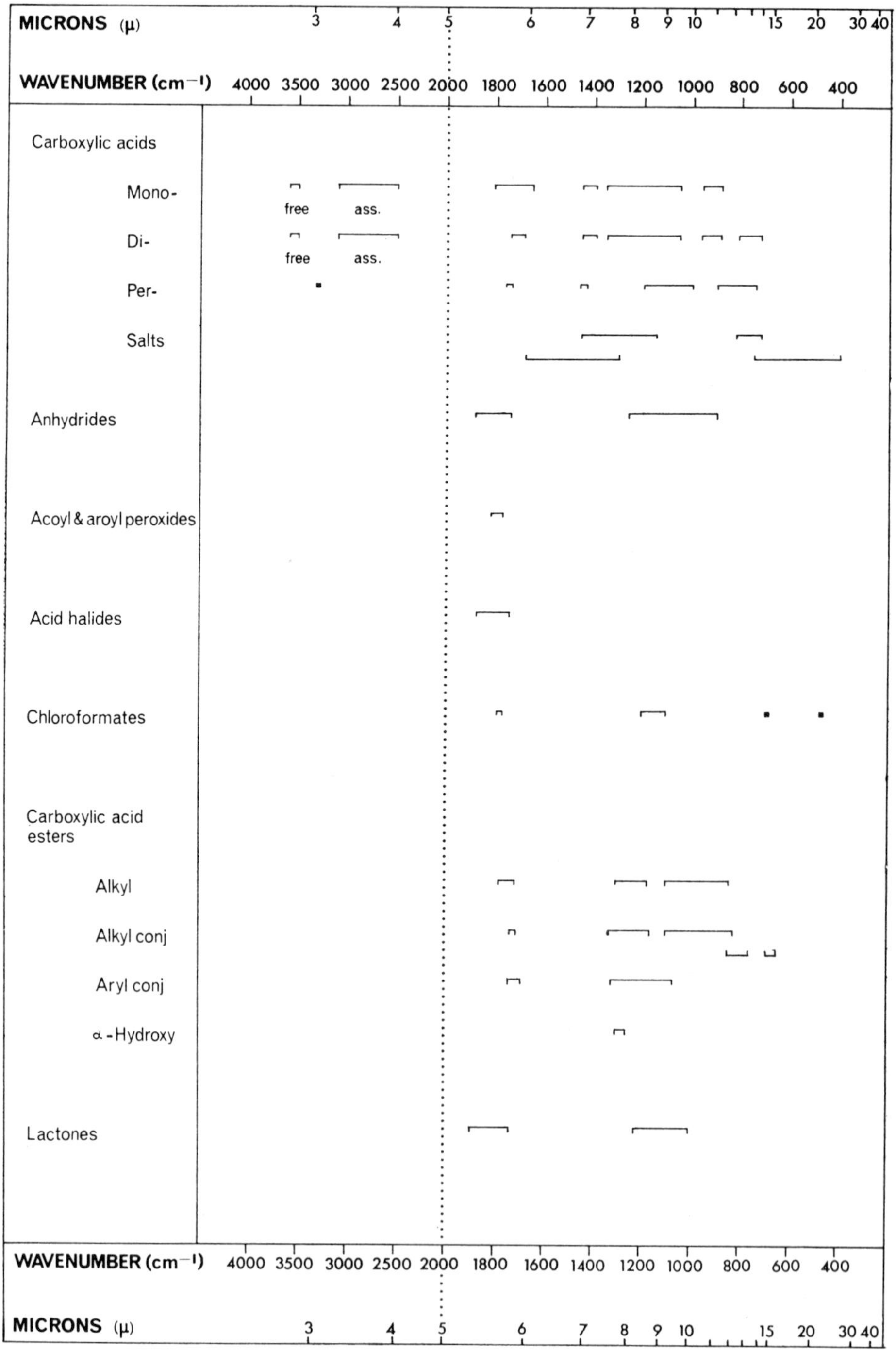

| MICRONS (μ) | | 3 | 4 | 5 | 6 | 7 | 8 | 9 | 10 | 15 | 20 | 30 40 |

| WAVENUMBER (cm⁻¹) | 4000 | 3500 | 3000 | 2500 | 2000 | 1800 | 1600 | 1400 | 1200 | 1000 | 800 | 600 | 400 |

Carboxylic acids

    Mono-   free   ass.

    Di-   free   ass.

    Per-

    Salts

Anhydrides

Acoyl & aroyl peroxides

Acid halides

Chloroformates

Carboxylic acid esters

    Alkyl

    Alkyl conj

    Aryl conj

    α -Hydroxy

Lactones

| WAVENUMBER (cm⁻¹) | 4000 | 3500 | 3000 | 2500 | 2000 | 1800 | 1600 | 1400 | 1200 | 1000 | 800 | 600 | 400 |

| MICRONS (μ) | | 3 | 4 | 5 | 6 | 7 | 8 | 9 | 10 | 15 | 20 | 30 40 |

NITROGEN COMPOUNDS
(excluding N-O compounds)

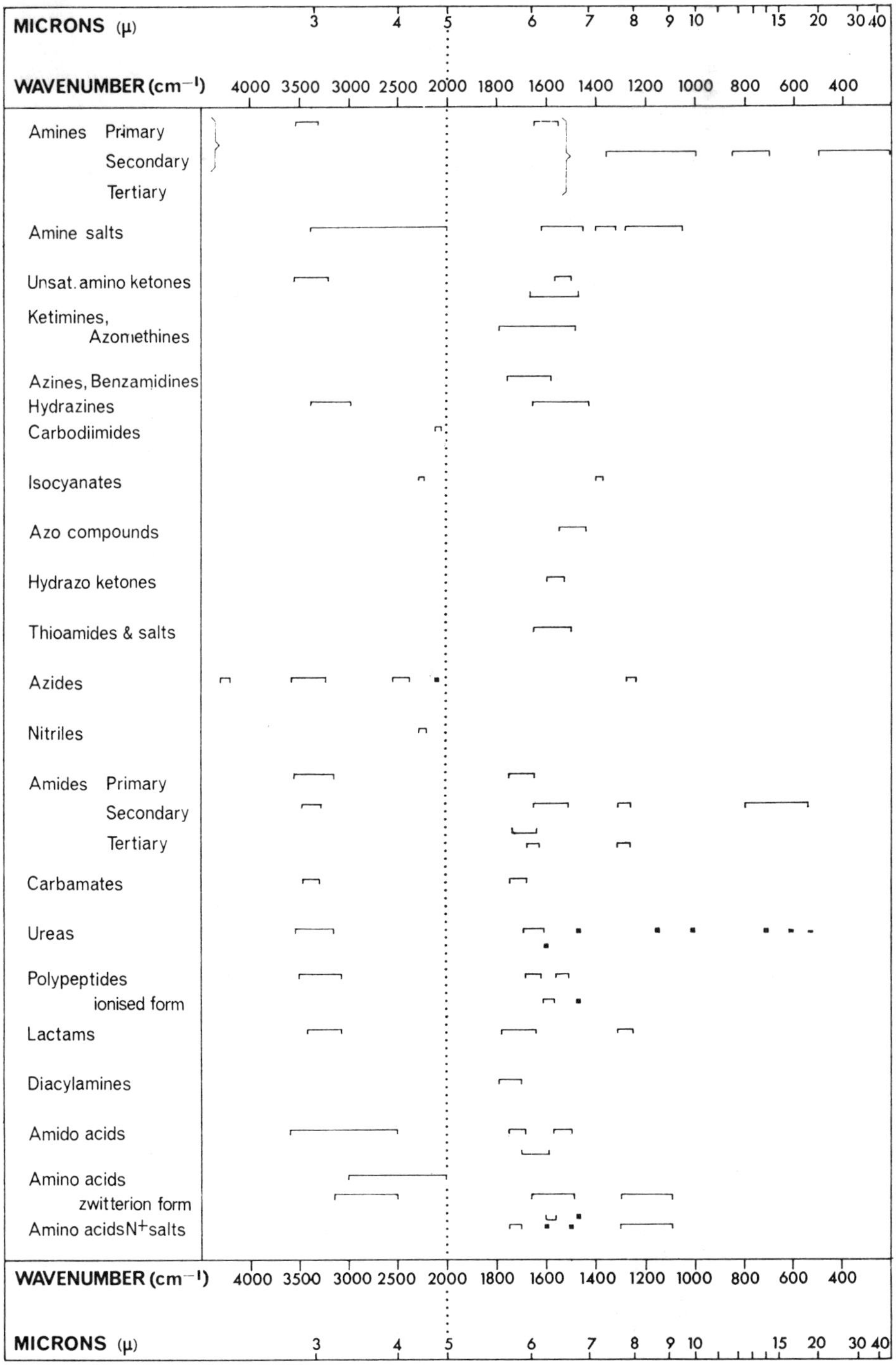

| MICRONS (μ) | 3 | 4 | 5 | 6 | 7 | 8 | 9 | 10 | 15 | 20 | 30 40 |

| WAVENUMBER (cm^{-1}) | 4000 | 3500 | 3000 | 2500 | 2000 | 1800 | 1600 | 1400 | 1200 | 1000 | 800 | 600 | 400 |

Amines   Primary
      Secondary
      Tertiary

Amine salts

Unsat. amino ketones

Ketimines, Azomethines

Azines, Benzamidines
Hydrazines
Carbodiimides

Isocyanates

Azo compounds

Hydrazo ketones

Thioamides & salts

Azides

Nitriles

Amides   Primary
      Secondary
      Tertiary

Carbamates

Ureas

Polypeptides
   ionised form

Lactams

Diacylamines

Amido acids

Amino acids
   zwitterion form
Amino acids N$^+$ salts

| WAVENUMBER (cm^{-1}) | 4000 | 3500 | 3000 | 2500 | 2000 | 1800 | 1600 | 1400 | 1200 | 1000 | 800 | 600 | 400 |

| MICRONS (μ) | 3 | 4 | 5 | 6 | 7 | 8 | 9 | 10 | 15 | 20 | 30 40 |

N-O COMPOUNDS

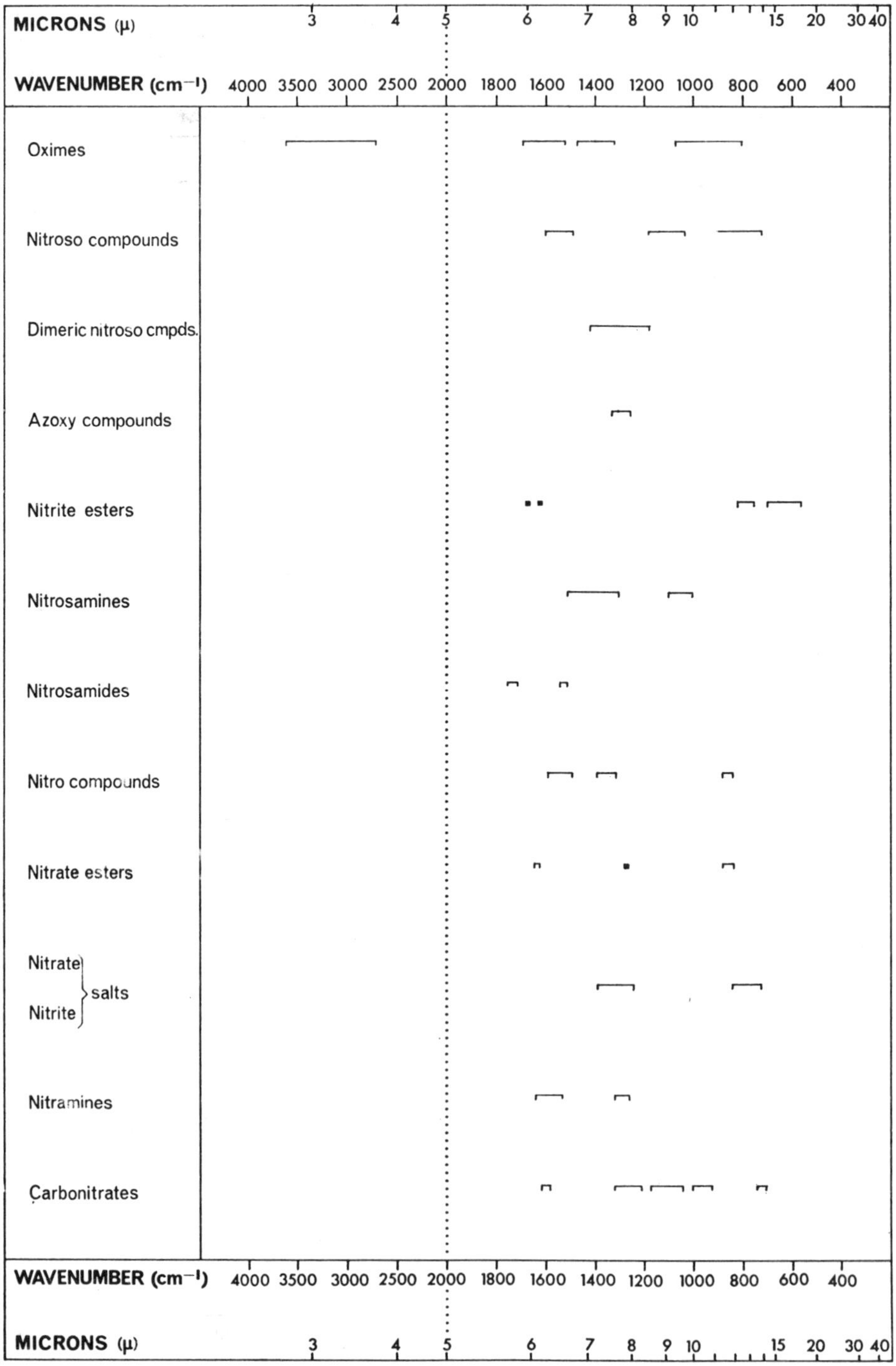

SULPHUR COMPOUNDS

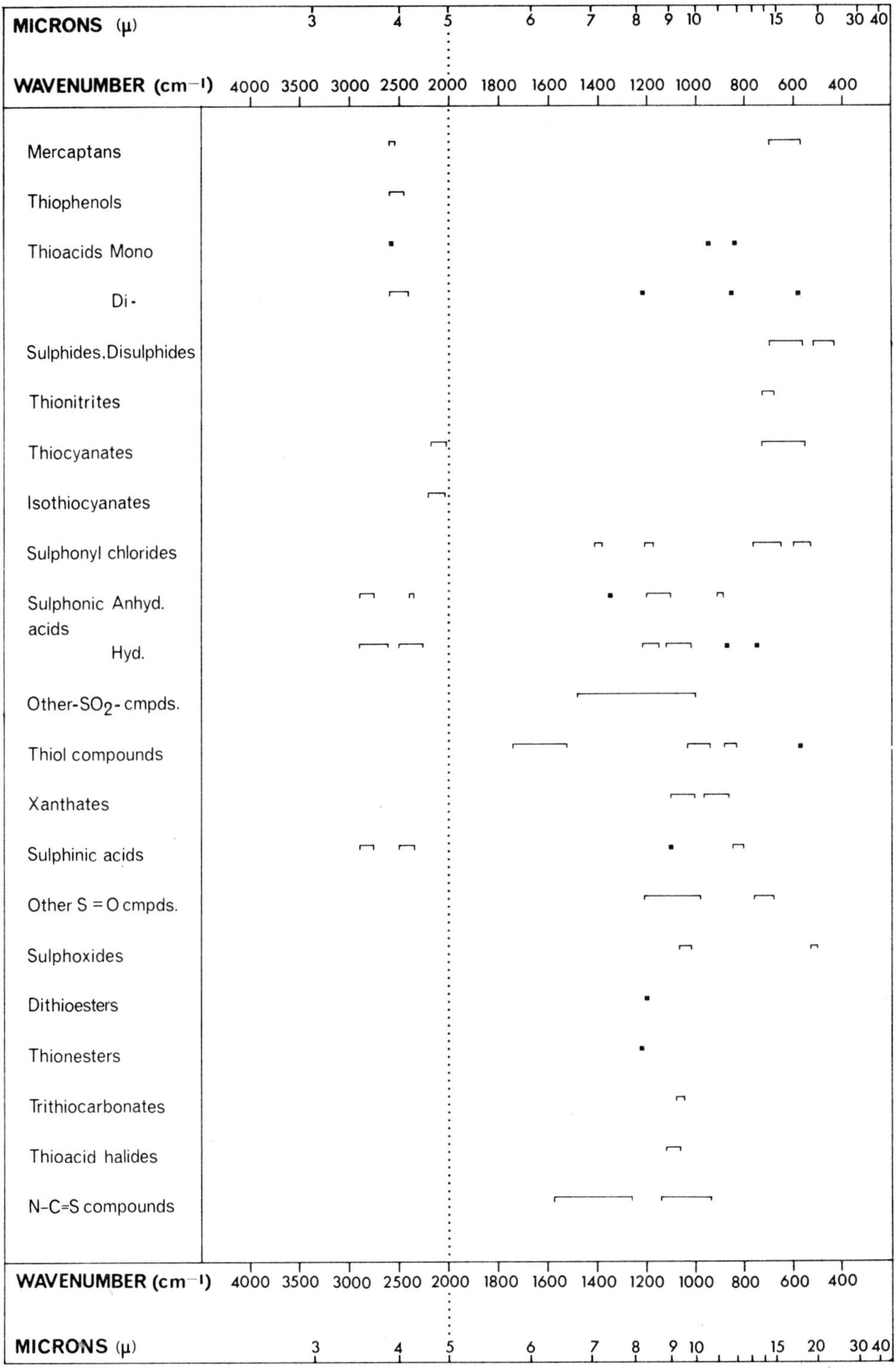
MICRONS (μ)
3 4 5 6 7 8 9 10 15 0 30 40
WAVENUMBER (cm⁻¹)
4000 3500 3000 2500 2000 1800 1600 1400 1200 1000 800 600 400
Mercaptans
Thiophenols
Thioacids Mono
Di-
Sulphides, Disulphides
Thionitrites
Thiocyanates
Isothiocyanates
Sulphonyl chlorides
Sulphonic acids Anhyd.
Hyd.
Other-SO2-cmpds.
Thiol compounds
Xanthates
Sulphinic acids
Other S=O cmpds.
Sulphoxides
Dithioesters
Thionesters
Trithiocarbonates
Thioacid halides
N-C=S compounds
WAVENUMBER (cm⁻¹)
4000 3500 3000 2500 2000 1800 1600 1400 1200 1000 800 600 400
MICRONS (μ)
3 4 5 6 7 8 9 10 15 20 30 40

SILICON COMPOUNDS

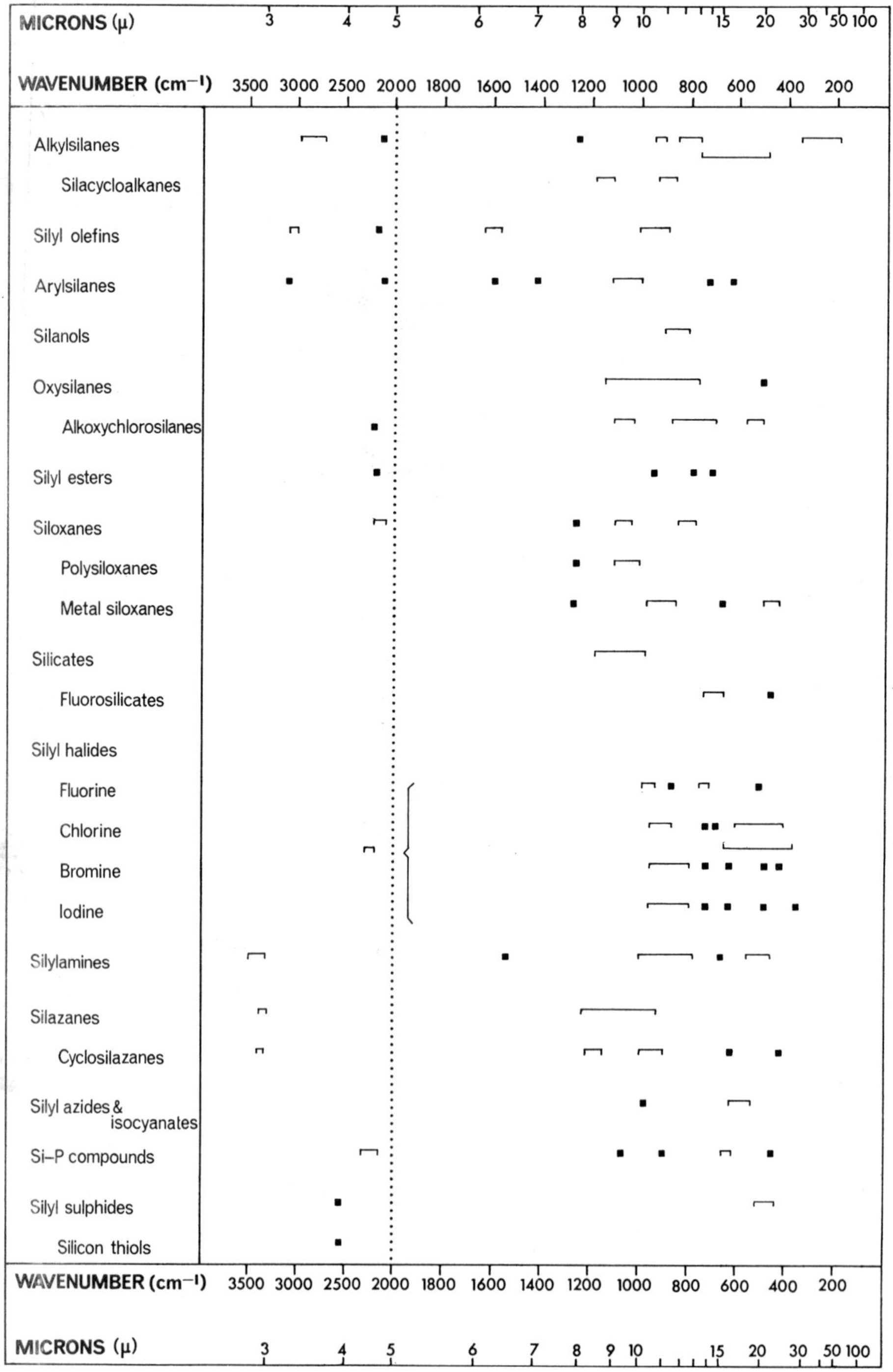

MICRONS (μ)   3   4   5   6   7   8   9   10   15   20   30 50 100

WAVENUMBER (cm⁻¹)   3500 3000 2500 2000 1800   1600 1400 1200 1000   800   600   400   200

Alkylsilanes
  Silacycloalkanes
Silyl olefins
Arylsilanes
Silanols
Oxysilanes
  Alkoxychlorosilanes
Silyl esters
Siloxanes
  Polysiloxanes
  Metal siloxanes
Silicates
  Fluorosilicates
Silyl halides
  Fluorine
  Chlorine
  Bromine
  Iodine
Silylamines
Silazanes
  Cyclosilazanes
Silyl azides & isocyanates
Si–P compounds
Silyl sulphides
  Silicon thiols

WAVENUMBER (cm⁻¹)   3500 3000 2500 2000 1800   1600 1400 1200 1000   800   600   400   200

MICRONS (μ)   3   4   5   6   7   8   9   10   15   20   30 50 100

Appendix D

INFRARED ABSORPTION FREQUENCIES OF FUNCTIONAL GROUPS*

Functional group	Absorption range (cm^{-1})	Example (cm^{-1})	Example compound
ALKANES			
a. linear			
CH_3 asymmetric	2970–2950	2967	n-Octane
CH_3 symmetric	2885–2865	2868	
CH_3 asymmetric	1465–1440	1466	
CH_3 symmetric	1380–1370	1380	
CH_2 asymmetric	2930–2915	2920	n-Octane
CH_2 symmetric	2860–2840	2854	
CH_2	1480–1450	1470	
$(CH_2)_n$ $n \geqslant 4$	723–720	723	n-Octane
$n = 3$	735–725	733	n-Pentane
$n = 2$	755–735	741	2-Methylpentane
$n = 1$	800–770	781	n-Propane
b. branched			
For CH_2, CH_3 wavenumbers, see section a.			
CH	2890	2890	Triphenylmethane
	1340	1341	
CH$_3$—CH—	1385–1380	1384	2-Methylheptane
│	1372–1366	1366	
CH$_3$	1175–1165		
	1160–1140		
	922–917		
CH$_3$	1395–1380	1393	2,2-Dimethylhexane
│	1375–1365	1366	
CH$_3$—C—	1252–1245		
│	1225–1195		
CH$_3$	930–925		
—C——C—	1165–1150	1160	3,4-Dimethylhexane
│ │	1130–1120	1122	
CH$_3$ CH$_3$	1080–1065	1071	

* Abbreviations: *sp* = sharp, *br* = broad, (w) = weak, (s) = strong.

Functional group	Absorption range (cm^{-1})	Example (cm^{-1})	Example compound
CH_3 | R—C—R | CH_3	1391–1381 1220–1190 1195–1185	1389 1192 1189	3,3-Dimethylhexane
C_2H_5 | R—CH | C_2H_5	1250 1150 1130	1250 1155 1131	3-Ethylhexane
C—C—C | CH_3	1160–1150		

c. cyclic compounds

Cyclopropane derivatives	3100–3072 3033–2995 1030–1000	3075 3028 1024	Cyclopropane
Cyclobutane derivatives	3000–2975 2924–2874 1000–960 or 930–890	2974 2896 901	Cyclobutane
Cyclopentane derivatives	2959–2952 2870–2853 1000–960 930–890	2951 2871 968 894	Cyclopentane
Cyclohexane derivatives	1055–1000 1015–950	1038 1014	Cyclohexane

For larger rings see section b.

UNSATURATED COMPOUNDS

a. isolated —C=C— bonds

CH_2=CH—	3095–3075 3030–2990 1648–1638 1420–1410 1000–980 915–905	3096 2994 1645 1420 994 912	1-Butene
CH_2=C<	3095–3075 1660–1640 1420–1410 895–885	3096 1661 1420 887	Methylpropene
—CH=C<	3040–3010 1680–1665 1350–1340 840–805	3037 1675 1351 812	3-Methyl-2-pentene

Functional group	Absorption range (cm^{-1})	Example (cm^{-1})	Example compound
C=C (cis) (with H, H substituents)	3040–3010	3030	cis-2-Butene
	1660–1640	1661	
	1420–1395	1406	
	730–675	675	
C=C (trans) (with H substituents)	3040–3010	3021	trans-2-Butene
	1700–1670	1701	
	1310–1295	1302	
	980–960	964	

b. conjugated —C=C— bonds

—C=C—C=C—	1629–1590	1592	1,3-Butadiene
	1820–1790	1821	

c. Allenic —C=C— bonds

—C=C=C—	1960–1940		
	1070–1060		

d. —C≡C— bonds

—C≡C—	2270–2250	2268	2-Pentyne

e. —C≡CH groups

CH (stretch)	3320–3300*	3320	1-Butyne
—C≡C—	2140–2100	2122	
CH (bend)	700–600		

AROMATIC COMPOUNDS
a. general

CH	3060–3010	
CH substitution bands, overtones	2000–1650 (w)	
C=C	1620–1590 sp	
	1590–1560 sp	
CH	1510–1480 sp	
	1450 sp	

b. mono-substitution

	1175–1125	1170	Toluene
	1110–1070	1088	
	1070–1000	1032	
	765–725	728 (s)	
	720–690	693 (s)	

* CCl$_4$ solutions only.

7

Functional group	Absorption range (cm^{-1})	Example (cm^{-1})	Example compound
c. di-substitution			
ortho	1225–1175	1185	*o*-Xylene
	1125–1090	1121	
	1070–1000	1053	
	765–735	741 (s)	
meta	1175–1125	1171	*m*-Xylene
	1110–1070	1095	
	1070–1000	1039	
	900–770	769 (s)	
	710–690	690 (s)	
para	1225–1175	1219	*p*-Xylene
	1125–1090	1120	
	1070–1000	1043	
	855–790	796 (s)	
d. tri-substitution			
1,2,3-	1175–1125	1162	1,2,3-Trimethylbenzene
	1110–1070	1095	
	1000–960	1009	
	800–755	765 (s)	
	740–695	710 (s)	
1,2,4-	1225–1175	1156	1,2,4-Trimethylbenzene
	1130–1090	1130	
	1000–960	1000	
	900–865	873 (s)	
	855–800	805 (s)	
1,3,5-	1175–1125	1165	1,3,5-Trimethylbenzene
	1070–1000	1039	
	860–810	836 (s)	
	705–685	690 (s)	
e. tetra-substitution			
1,2,4,5-	870–855	870	1,2,4,5-Tetramethyl-benzene

ALCOHOLS
a. general

OH unbridged group	3650–3590 *sp*		
OH inter- and intra- molecularly H-bonded	3570–3450		
OH intermolecularly H-bonded	3400–3200 *br*		

b. primary alcohols

	1350–1260	1339	1-Pentanol
	1065–1020	1028	

Functional group	Absorption range (cm^{-1})	Example (cm^{-1})	Example compound
c. secondary alcohols			
	1370–1260	1369	2-Pentanol
	1120–1080	1111	
d. tertiary alcohols			
	1410–1310	1379	2-Methylbutanol-2
	1170–1120	1124	
e. aromatic ring hydroxy compounds			
OH unbridged	3617–3599 *sp*		
OH dimer	3460–3322 *br*		
OH polymer	3370–3322 *br*		
	1410–1310	1350	Phenol
	1225–1175	1225	
PEROXIDES			
a. aliphatic			
	1820–1810		
	1800–1780		
	890–820		
b. aromatic			
	1805–1780		
	1785–1755		
	1020–980		
ETHERS			
a. aliphatic			
O—CH$_3$	2830–2815		
C—O—C	1150–1060	1140	Diethyl ether
O—(CH$_2$)$_4$	742–734		
O—CH$_3$	1455		
b. aromatic			
=C—O—C	1275–1200	1247	Anisol
C—O—C	1075–1020	1038	
c. cyclic			
C—O—C	1140–1070		
d. epoxides			
	1260–1240	1261	1;2-Epoxybutane
trans compounds	890		
cis compounds	830	826	1;2-Epoxybutane

Functional group	Absorption range (cm^{-1})	Example (cm^{-1})	Example compound
e. tetrahydrofuran derivatives			
	1098–1075	1076	Tetrahydrofuran
	915–913	912	
f. trioxans			
	1175	1172	Trioxan
	958	957	
g. tetrahydropyran derivatives			
	1120–1080		
	1100–900		
	825–805		
h. dioxan derivatives			
	1125	1122	Dioxan

KETALS, ACETALS

R   O—C
  \ /
   C
  / \
R   O—C

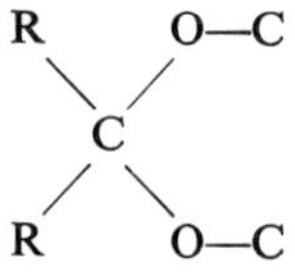	1190–1158		
	1143–1124		
	1098–1063		

KETONES

a. aliphatic

	1725–1705	1727	Butanone
	1325–1215	1269	
	1200	1215	

b. unsaturated

C=C	1650–1620	1618	Methyl vinyl ketone
C=O	1685–1665	1684	

c. aromatic

Aryl, alkyl	1700–1680	1694	Acetophenone
Aryl, aryl	1670–1660		

d. cyclic

4- and 5-membered rings	1775–1740	1739	Cyclopentanone
6- and 7-membered rings	1725–1700	1703	Cycloheptanone

e. diketones

α-Diketones	1730–1710	1721	Diacyl ketone
β-Diketones	1640–1540		
γ-Diketones	1725–1705		

Functional group	Absorption range (cm^{-1})	Example (cm^{-1})	Example compound

f. halogen substituted

α,α-Dihalogen substituted	1765–1745		
α-Halogen substituted	1745–1725		

ALDEHYDES
a. general

	2900–2700 (2 bands)		
CH	2720–2700		
	975–780		

b. aliphatic

C=O	1740–1720	1735	Butyraldehyde
CH	1440–1325	1390	

c. unsaturated

C=O	1650–1620	1637	Crotonaldehyde
C=O α, β unsaturated	1690–1650		

d. aromatic

CH	2750–2720 *sp*	2725	Benzaldehyde
C=O	1715–1695	1701	
	1415–1350	1391	
	1320–1260	1312	
	1230–1160	1203	

CARBOXYLIC ACIDS
a. general

OH	3200–2500 *br*		
CH	1440–1396		
	1320–1210		
OH dimer	950–900 *br*		
C=O halogen substituted	1740–1720		
C=O aliphatic	1720–1700	1718	Acetic acid
C=O unsaturated	1710–1690	1698	Crotonic acid
C=O aromatic	1700–1680	1695	Benzoic acid
C=C unsaturated	1660–1620	1655	Crotonic acid

b. carboxylic ions

C=O	1610–1560		
C=O	1420–1300		

KETENES

	2155–2140	2155	Ketene
	1135–1120	1136	

Functional group	Absorption range (cm^{-1})	Example (cm^{-1})	Example compound
ESTERS			
C=O unsaturated, aryl	1800–1770		
C=C unsaturated, aryl	1730–1710	1718	
C—O acrylates, fumarates	1300–1200	1282	Ethyl acrylate
C—O	1190–1130	1192	
C=O electronegatively substituted	1770–1745		
C=O α, γ keto	1755–1740		
C=O saturated	1750–1735	1744	Methyl acetate
C=O β keto	1660–1640		
C—O benzoates, phthalates	1310–1250 1150–1100	1277 1108	Methyl benzoate
C—O acetates	1250–1230 1060–1000	1246 1047	Propyl acetate
C—O phenolic acetates	1205		
C—O formate	1200–1180	1190	Propyl formate
LACTONES			
β-Lactones	1840–1800		
γ-Lactones	1780–1760	1776	Butyrolactone
δ-Lactones	1750–1730 1280–1150	1168	
ANHYDRIDES			
a. aliphatic			
C=O	1850–1800	1842	Acetic acid anhydride
C=O	1785–1760	1783	
C—O	1170–1050	1134	
b. aromatic			
C=O	1880–1840	1866	Phthalic acid anhydride
C=O	1790–1770	1773	
C—O	1300–1200	1267	
c. cyclic			
C=O	1870–1820	1818	Glutaric acid anhydride
C=O	1800–1750	1772	
ACID CHLORIDES			
a. aliphatic			
C=O	1815–1770	1802	Acetyl chloride

Functional group	Absorption range (cm^{-1})	Example (cm^{-1})	Example compound
b. aromatic			
C=O	1775–1760	1773	Benzoyl chloride
C=O	1730–1700	1726	
AMIDES			
a. primary			
NH free	3500		
NH free	3400		
NH bridged	3350	3346	Butyramide
NH bridged	3190	3191	
C=O	1660–1640	1660	
	1430–1400	1430	
b. secondary			
NH free *trans*	3460–3400		
NH free *cis*	3440–3420		
NH bridged *trans*	3320–3270	3280	*N*-Methylacetamide
NH bridged *cis*	3180–3140		
bridged *cis, trans*	3100–3070	3090	
C=O	1680–1630	1652	
NH	1570–1510	1564	
	720 *br*	725	
c. tertiary			
C=O	1670–1630	1670	*N,N*-Dimethylformamide
AMINO ACIDS			
NH	3130–3030 *br*		
	2760–2530 (not always present)		
	2140–2080		
C=O	1720–1680		
Ionised form	1600–1560		
	1300		
C=O α-amino acids	1754–1720		
C=O β, γ-amino acids	1730–1700		
Amino acid hydrochlorides	3030–2500 (more bands)		
NH amino acid hydro-chlorides	1660–1590		
NH amino acid hydro-chlorides	1550–1490		
AMINES			
a. general			
N—CH$_3$	2820–2730		
N—CH$_3$	1426		
C—N	1410		

Functional group	Absorption range (cm^{-1})	Example (cm^{-1})	Example compound

b. aliphatic, primary

NH free	3500–3200	3350	Ethylamine
	(2 bands)	3210	
NH	1650–1590	1630	
	1200–1150		
	1120–1030	1100	

c. aliphatic, secondary

NH free	3500–3200	3230	Dipropylamine
	(1 band)		
NH	1650–1550		
C—N	1200–1120	1126	
C—N	1150–1080	1090	

d. aliphatic, tertiary

| C—N | 1230–1130 | 1175 | Ethyldimethylamine |
| C—N | 1130–1030 | 1070 | |

e. aromatic, primary

	3510–3450	3460	Aniline
	3420–3380	3413	
	1630–1600	1621	

f. aromatic, secondary

| Free | 3450–3430 | | |
| Bridged | 3400–3300 | 3400 | N-Methylaniline |

UNSATURATED NITROGEN COMPOUNDS
a. imines

| NH | 3400–3300 | | |
| C=N | 1690–1640 | | |

b. oximes

Liquid	3602–3590		
Solid	3250		
Solid	3115		

| Aliphatic | 1680–1665 | | |

Aromatic	1650–1620		
	1300		
	900		

Functional group	Absorption range (cm^{-1})	Example (cm^{-1})	Example compound
CYANIDES, ISOCYANIDES			
C≡N unconjugated	2265–2240	2256	Ethyl cyanide
C≡N conjugated or aromatic	2240–2220	2222	Benzyl cyanide
C≡N cyanide, thiocyanide complex	2200–2000		
N≡C alkyl isocyanide	2183–2150	2166	Methyl isocyanide
N≡C aryl isocyanide	2140–2080	2100	Phenyl isocyanide
CYCLIC NITROGEN COMPOUNDS			
a. pyridines, quinolines			
CH	3100–3000	3030	Pyridine
C=C, C=N	1615–1590	1590	
	1585–1550		
	1520–1465	1490	
	1440–1410		
	920–690	707	
	(substituent dependent)		
b. pyrimidines			
CH	3060–3010		
C=C, C=N	1580–1520		
Ring	1000–900		
UNSATURATED NITROGEN-NITROGEN COMPOUNDS			
Azo compounds	1630–1575		
N=N azides	2160–2120	2130	Phenylazide
N=N azides	1340–1180	1297	
NITRO COMPOUNDS			
a. aliphatic			
	1570–1500	1546	2-Nitrobutane
	1385–1365	1362	
	880	879	
b. aromatic			
	1550–1510	1527	Nitrobenzene
	1370–1330	1351	
	849	853	

Functional group	Absorption range (cm^{-1})	Example (cm^{-1})	Example compound

PHOSPHORUS COMPOUNDS

O—H phosphoric acids	2700–2560 *br*		
P—H	2440–2350 *sp*		
P=O	1350–1250		
P=O	1250–1150		
P—O—C	1240–1190		
P—O—R	1190		
P—O—C	1170–1150		
P—O—C	1050–990		
P—O—P	970–940		
P—F	885		
P=S	840–600		
O—P—H	865–840		
O—P—O	590–520		
O—P—O	460–440		

PHOSPHORUS-CARBON COMPOUNDS

P—C aromatic	1450–1435		
P—C aliphatic	1320–1280	1298	Trimethylphosphine
P—C	750–650	707	
PO$_4^{3-}$ aryl phosphates	1080–1040		
PO$_4^{3-}$ alkyl phosphates	1180–1150		
PO$_4^{3-}$ alkyl phosphates	1080		

DEUTERATED COMPOUNDS

O—D deuterated alcohols	2650–2400		
O—D deuterated carboxylic acids	675		

SULPHUR COMPOUNDS

C=S	1400–1300	1357	Dithioacetic acid
S=S	1200–1050		
P=S	840–600		
SH mercaptans	2600–2550	2580	Ethyl mercaptan
C—S mercaptans	700–600	665	
C—S—C dialkyl sulphides	750–600	726	Methyl ethyl sulphide
	710–570	676	
	660–630	654	
Aliphatic sulphones	1410–1390	1407	Dimethylsulphone
	1350–1300	1316	
Sulphonic acids	1210–1150		
	1060–1030		
	650		
S—CH$_3$	1325		

Functional group	Absorption range (cm⁻¹)	Example (cm⁻¹)	Example compound

SILICON COMPOUNDS

Functional group	Absorption range (cm⁻¹)	Example (cm⁻¹)	Example compound
SiH alkylsilanes	2300–2100	2175	Dimethylsilane
Si(CH₃)₂	1265–1258	1262	
	814–800		
	800		
Si(CH₃)₃	1260–1240	1259	Methoxytrimethylsilane
	850–830	844	
	760	763	
Si—C aromatic	1429		
	1130–1090		
Si—C	860–715		
Si—O siloxanes	1100–1000		
Si—O—C open-chain	1090–1020		
Si—O—Si open-chain	1097		
Si—O—Si cyclic	1080–1010		

HALOGEN COMPOUNDS

a. iodine compounds

ca. 500

b. bromine compounds

700–500

c. chlorine compounds

Monochloro	800–600		
	750–700		
Fully chlorinated compounds	780–710		

d. fluorine compounds

1400–1000
1100–1000

Fully fluorinated compounds   745–730

INORGANIC COMPOUNDS

a. sulphates

	1200–1140	1143	Potassium sulphate
	1130–1080	1117	
	680–610	617	

b. nitrates

	1380–1350	1370	Potassium nitrate
	840–815	825	

c. nitrites

840–800
750

Functional group	Absorption range (cm^{-1})	Example (cm^{-1})	Example compound

d. water of crystallisation
	1630–1615		

e. halogen-oxygen salts
Chlorates	980–930	978	Potassium chlorate
	930–910	932	
Bromates	810–790	793	Potassium bromate
Iodates	785–730	756	Potassium iodate

f. carbonates
	1450–1410	1410	Calcium carbonate
	880–860	875	

g. selenium compounds
Selenates	895
	420
Selenites	740
	460

Appendix E

GENERAL BIBLIOGRAPHY

BOOKS

H. C. Allen and P. C. Cross, *Molecular Vib-Rotors*, John Wiley & Sons, New York, 1963, 320 pp.

N. L. Alpert, W. E. Keiser and H. A. Szymanski, *Theory and Practice of Infrared Spectroscopy*, 2nd edn., Plenum Press, New York (Heyden & Son, London) 1970, 380 pp.

C. N. Banwell, *Fundamentals of Molecular Spectroscopy*, McGraw-Hill, New York, 1966, 282 pp.

G. M. Barrow, *The Structure of Molecules*, Benjamin, New York, 1963, 156 pp.

G. M. Barrow, *Introduction to Molecular Spectroscopy*, McGraw-Hill, New York, 1962, 332 pp.

N. B. Colthup, L. H. Daly and S. E. Wiberley, *Introduction to Infrared and Raman Spectroscopy*, Academic Press, New York, 1964, 511 pp.

R. T. Conley, *Infrared Spectroscopy*, Allyn & Bacon, Boston, 1966, 293 pp.

B. W. Cook and K. Jones, *A Programmed Introduction to Infrared Spectroscopy*, Heyden & Son, London, 1972.

A. D. Cross, *An Introduction to Practical Infrared Spectroscopy*, 3rd edn., Butterworths, London, 1969, 110 pp.

J. R. Ferraro, *Low Frequency Vibrations of Inorganic and Coordination Compounds*, Plenum Press, New York (Heyden & Son, London) 1971, 309 pp.

Ian Flemming and D. H. Williams, *Spectroscopic Methods in Organic Chemistry*, McGraw-Hill, London, 1966, 215 pp.

L. A. Gribov, *Intensity Theory for Infrared Spectra of Polyatomic Molecules*, Plenum, New York, 1964, 120 pp.

M. L. Hair, *Infrared Spectroscopy in Surface Chemistry*, Dekker, New York, 1967, 314 pp.

D. N. Kendall, *Applied Infrared Spectroscopy*, Reinhold, London, 1966, 560 pp.

K. E. Lawson, *Infrared Absorption of Inorganic Substances*, Reinhold, London, 1961, 227 pp.

L. H. Little, *Infrared Spectra of Absorbed Species*, Academic Press, London, 1966, 428 pp.

C. E. Meloan, *Elementary Infrared Spectroscopy*, McMillan, London, 1963, 180 pp.

K. Nakamoto, *Infrared Spectra of Inorganic and Coordination Compounds*, John Wiley & Sons, New York, 1963, 330 pp.

W. J. Potts Jr, *Chemical Infrared Spectroscopy*, Vol. 1: Techniques, John Wiley & Sons, New York, 1963, 322 pp.

H. Siebert, *Anwendung der Schwingungsspektroskopie in der Anorganischen Chemie*, Springer Verlag, Berlin, 1966, 209 pp.

W. W. Wendlandt and H. G. Hecht, *Reflectance Spectroscopy*, Interscience, New York, 1966, 275 pp.

R. G. White, *Handbook of Industrial Infrared Analysis*, Plenum Press, New York, 1964, 452 pp.

R. Zbinden, *Infrared Spectroscopy of High Polymers*, Academic Press, New York, 1964, 260 pp.

Index

Further Heyden Publications
on Infrared Spectroscopy

IRSCOT SYSTEM
by R. G. J. Miller and H. A. Willis

Currently covering ten chemical classes, the popular and widely renowned IRSCOT system is a quick and simple-to-use reference guide consisting of correlation tables, a master index and data cards providing telegram-style information on specific infrared bands.

INFRARED VAPOUR SPECTRA
by D. Welti

This important work shows why vapour spectra are as readily applicable to structural determination as liquid or solid spectra. The author's lucid description of relevant techniques and instrumentation is accompanied by over 300 fully indexed vapour spectra.

INFRARED ANALYSIS OF ESSENTIAL OILS
by J. Bellanato and A. Hidalgo

This is a study of the application of infrared spectroscopy to the characterisation of essential oils and their constituents. It covers 35 essential oils, and the text is illustrated by over 200 spectra of essential oils and their constituent compounds.

LABORATORY METHODS IN INFRARED SPECTROSCOPY (2ND EDN.)
by R. G. J. Miller and B. C. Stace

First published in 1965, this book rapidly established itself as a standard reference work for all who use infrared spectroscopy. In this revised and greatly enlarged second edition, eminent spectroscopists acquaint you with a wealth of short cuts and 'tricks of the trade' that have saved them hours of needless experimentation.

A PROGRAMMED INTRODUCTION TO INFRARED SPECTROSCOPY
by B. W. Cook and K. Jones

In an exciting and logical manner, this unique publication instructs students and technicians in the theory and practice of infrared spectroscopy. The reader, working at his own pace, tackles the subject step-by-step — absorbing each concept before progressing to the next — until he is ultimately able to prepare a sample, set-up and operate a spectrophotometer and interpret the resulting spectrogram.

For more details on publications on infrared spectroscopy write to :
Heyden & Son Ltd., Spectrum House, Alderton Crescent, London NW4 3XX
Heyden & Son GmbH., Steinfurter Strasse 45, 4440 Rheine/Westf., Germany